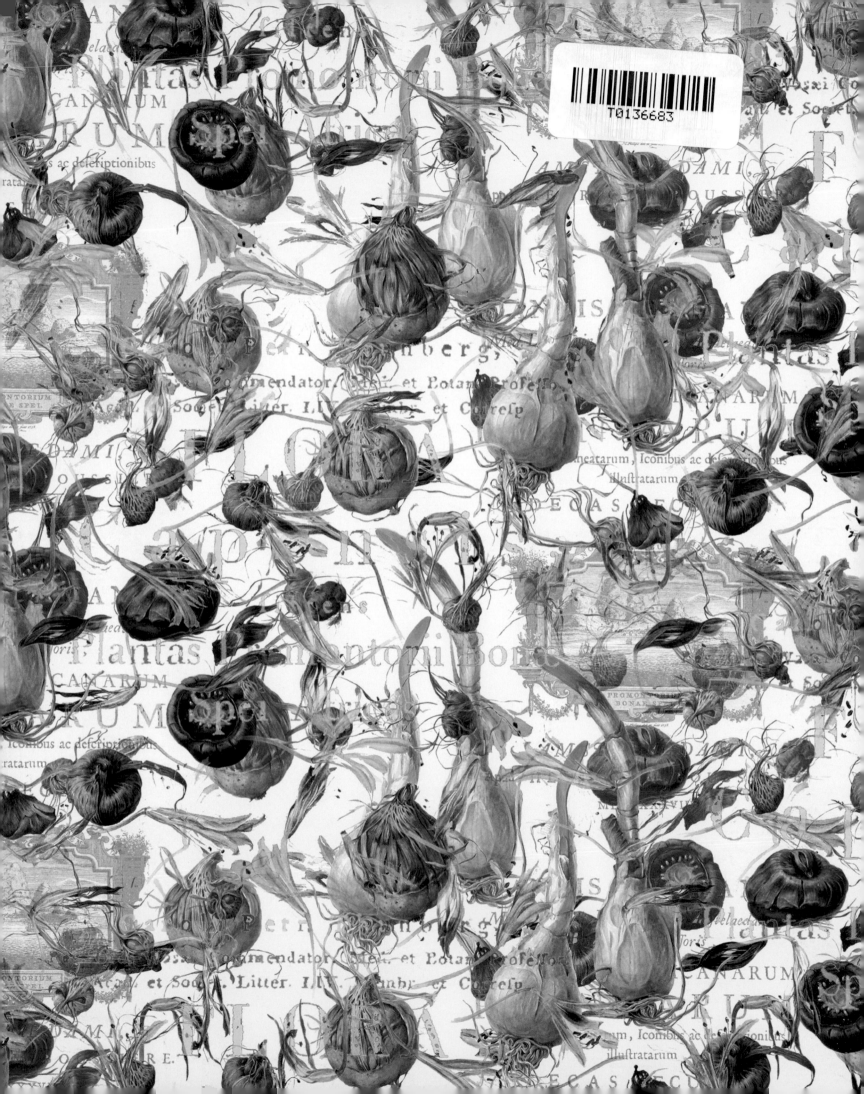

The SMALLEST KINGDOM

The
SMALLEST

MIKE AND LIZ FRASER

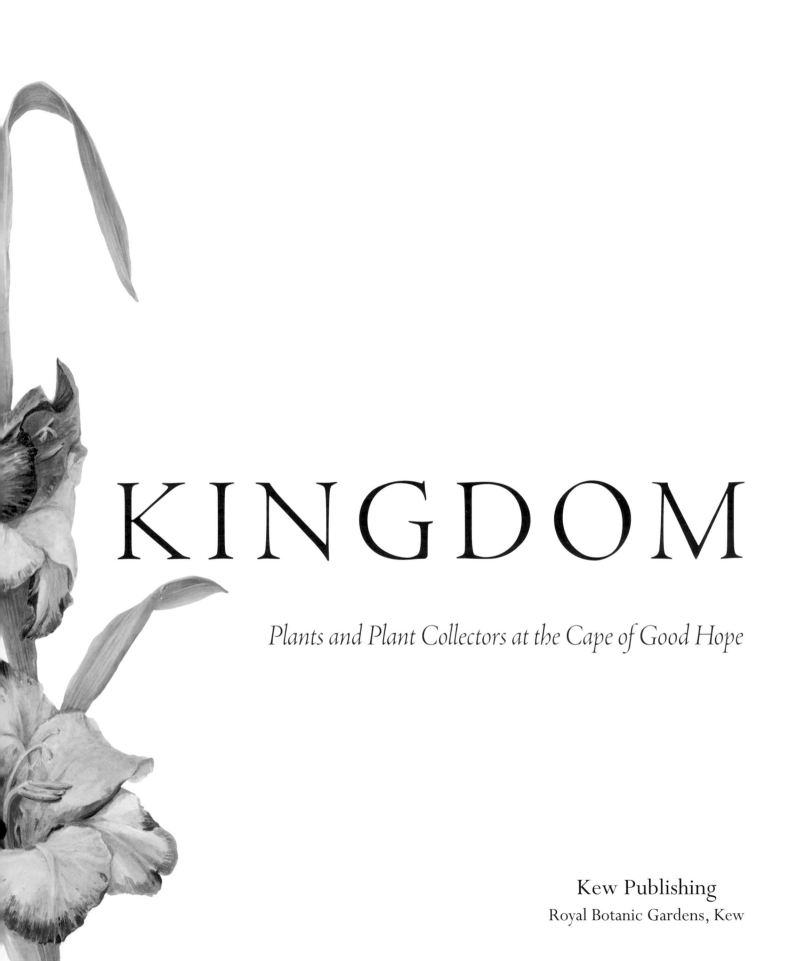

KINGDOM

Plants and Plant Collectors at the Cape of Good Hope

Kew Publishing
Royal Botanic Gardens, Kew

Contents

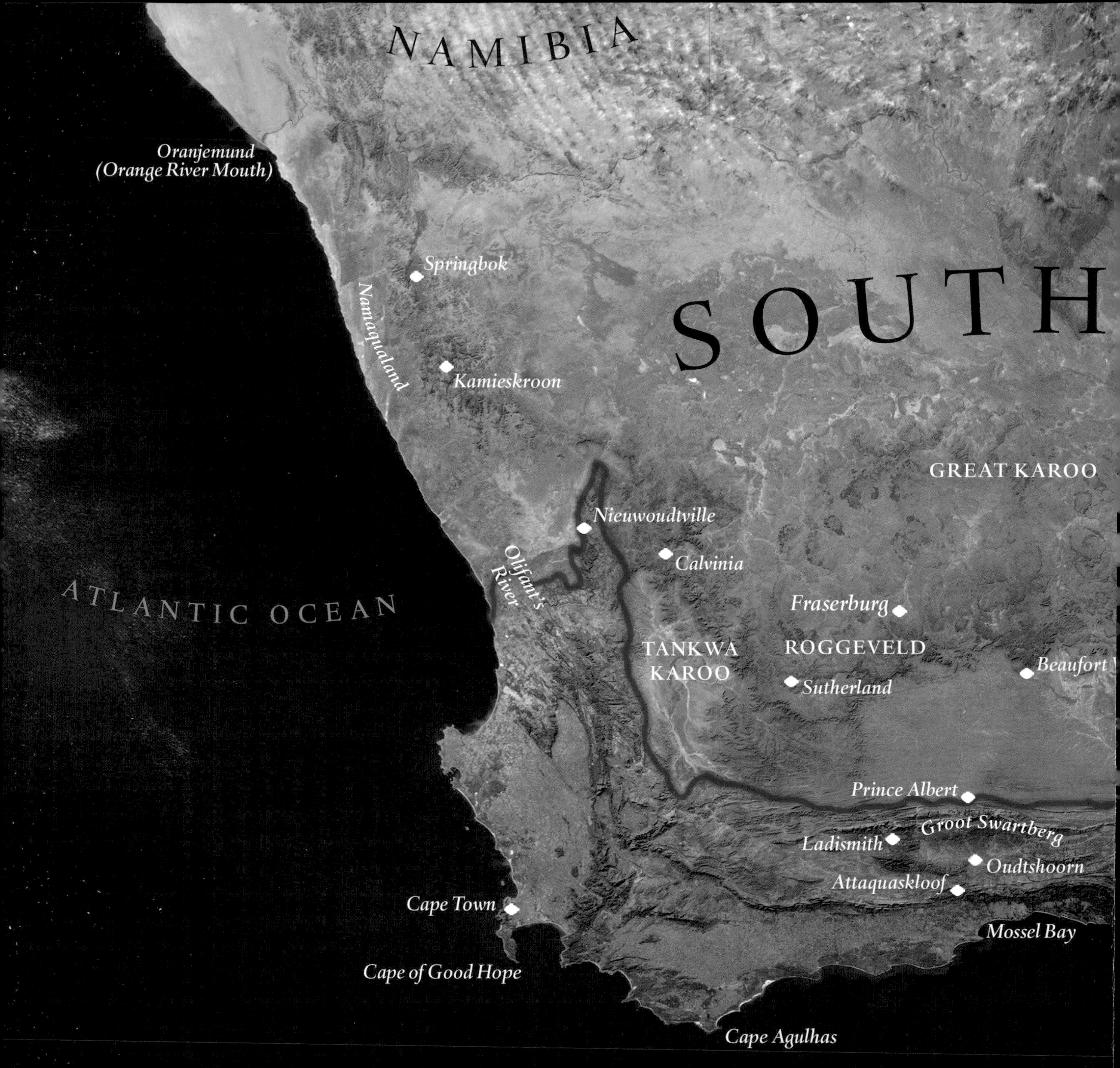
NAMIBIA
Oranjemund
(Orange River Mouth)
Springbok
Namaqualand
Kamieskroon
SOUTH
GREAT KAROO
Nieuwoudtville
Calvinia
Olifant's River
ATLANTIC OCEAN
Fraserburg
TANKWA KAROO
ROGGEVELD
Sutherland
Beaufort
Prince Albert
Groot Swartberg
Ladismith
Oudtshoorn
Attaquaskloof
Cape Town
Mossel Bay
Cape of Good Hope
Cape Agulhas

Locating The Smallest Kingdom

The Cape from space. The narrow limits of the Cape Floral Kingdom are neatly bounded by the sea to the south and west, and inland by the distinctive rocky ridges of the Cape Folded Belt mountains running north and east from Cape Town. The most conspicuous features of the modern landscape, these mountains were formed 300 million years ago from deep sandstone deposits that were lifted and buckled by massive tectonic movements of the Earth's crust. By a curious quirk of geological fate, Table Mountain escaped this buckling and retains its flat shape and horizontal stratification as text-book evidence of its sedimentary origins.

The upward-thrusting peaks of the mountains were gradually exposed by the erosion of softer shales lying on top of the sandstone sediments. The remaining shales now make up the fertile, rolling 'downs' in the valleys.

BOTSWANA

AFRICA

Durban

INDIAN OCEAN

Aberdeen   Somerset East

East London

Grahamstown

Sundays River

iaanskloof

Algoa Bay   Kwaaihoek

Port Elizabeth

Cape St Francis

While the very hard, slowly-eroding sandstone plugs of the Cape Folded Belt mountains have remained much the same over the last 65 million years, the Cape's coastal lowlands have experienced much greater change. Here, the rise and fall in sea level over the last two million years has varied their extent from almost non-existent to three times their present size. For much of this time a drop in sea level of about 6 m saw the west coast extending seaward by about 15 km beyond its present limit. *NASA/courtesy of nasaimages.org*

youthful botanical ignorance we knew as 'geraniums', grew wild like weeds, while elegant gladioli and bold blue agapanthus popped up amongst ubiquitous reeds, heaths and scrubby bushes. A suspicion that these had been surreptitiously planted by guerrilla gardeners was soon dispelled by the realisation that the Cape is the original home of a multitude of cultivated flowers that today are denizens of garden and greenhouse.

Having planned to stay at the Cape for one year, 12 flashed blissfully by before we finally prised ourselves away and migrated northwards with the swallows. Helping to soften the blow of being back in Britain was the discovery of nurseries and gardens full of familiar Cape flowers. The ancestors of these had been collected by a succession of plant hunters over three centuries and in whose footsteps we had waveringly followed.

Often as colourful and interesting as their plants, some of the botanical explorers also left detailed and fascinating accounts of their time at the Cape, their journals providing valuable historical insight into a unique landscape and its original and colonial occupants. Many of them also represented the great institutions of Europe, including the Royal Botanic Gardens, Kew, whose first dedicated international plant collector, Francis Masson, was dispatched to the Cape in 1772. The plants that he and others sent back became the subjects of intense scientific scrutiny by academics and, perhaps even more enthusiastically, experimentation and development by the rapidly growing horticultural industry. The profusion of varieties that were bred and marketed reflected the popularity of Cape plants and the skill and industry of the plantsmen of the day.

The diversity of unusual species shipped north in the eighteenth and nineteenth centuries also alerted scientists to the fact that an extraordinary flora lay at the Cape. Here was a botanical treasurehouse ablaze with thousands of species, many of them extremely rare and found nowhere else, whose richness and diversity exceeded all other temperate and the majority of tropical regions.

The Smallest Kingdom is our celebration of Cape plants and the people who discovered and developed many of them into today's horticultural gems. It is by necessity, but unapologetically, an Anglocentric (or Scottishcentric if there was such a word) attempt to give recognition and prominence to the Cape's fabulous flora and its outstanding contribution to the gardens of the world.

ACKNOWLEDGEMENTS

Our time at the Cape was made possible in the first instance thanks to the good offices of Robert Prŷs-Jones, formerly of the Percy Fitzpatrick Institute of Ornithology at the University of Cape Town (UCT) where MF undertook post-graduate research supported by a J. W. Jagger Overseas Student Scholarship.

Derek Clark set us off on the right track. With patience and good humour he introduced us to the many splendours, particularly botanical, of our study site at the Cape of Good Hope Nature Reserve where he was Senior Ranger. His enthusiasm was infectious, and soon infected us.

Friends and colleagues at UCT and other institutions, and members of the Cape Bird Club and the Botanical Society of South Africa, made us welcome and introduced us to the local wildlife over the years. For which, our thanks go to Nigel and Claudia Adams, Alice Ashwell, Mark Beeston, Aldo Berruti Callan Cohen, John Cooper, Richard and Sue Dean, Diana and John Deeks, Atherton de Villiers, Greg Forsyth, John Graham, Tony and Jen Grogan, Phil Hockey, Jenny Jarvis, Richard Knight, Howard and Toni Langley, Belle Leon, Dave Macdonald, Rob

opposite
Foliage, Flowers, and Fruit of the South African Silver Tree
Painting by Marianne North. 'On the right below is a head of female flowers, and above on the left a ripe cone … in the middle and above are heads of male flowers.'
Official Guide to the North Gallery.

Short-horned grasshopper

BOTSWANA

AFRICA

◆ Durban

INDIAN OCEAN

◆ Aberdeen   ◆ Somerset East

◆ East London

Grahamstown ◆

Sundays River

...anskloof

Algoa Bay   ◆ Kwaaihoek

Port Elizabeth

Cape St Francis

While the very hard, slowly-eroding sandstone plugs of the Cape Folded Belt mountains have remained much the same over the last 65 million years, the Cape's coastal lowlands have experienced much greater change. Here, the rise and fall in sea level over the last two million years has varied their extent from almost non-existent to three times their present size. For much of this time a drop in sea level of about 6 m saw the west coast extending seaward by about 15 km beyond its present limit.   *NASA/courtesy of nasaimages.org*

Erica heaths are abundant at the Cape and are a major component of the fynbos ('*fayn-boss*') vegetation that characterises the region. Left to right, *E. sessiliflora*, *E. lanuginosa*, and *E. coccinea*.

Introduction & Acknowledgements

At first no object attracted my notice till I had sufficiently admired the majestic amphitheatre of the mountains, in which the town reposes. Every thing wore, to my eye at least, a pleasing aspect: it was the charm of novelty which cast an agreeable hue over the whole scene; even the smallest object interested me, and whatever I beheld seemed to present itself as a subject for my future investigation. On the first arriving at a foreign country, there is a sensation so delightful and so peculiar to an inquisitive mind, that language can convey but little of it to a reader.

William Burchell, *Travels in the Interior of Southern Africa*, 1822.

Visitors to Cape Town before and after Burchell have been captivated by its 'pleasing aspect' and, not least, 'the majestic amphitheatre' created by Table Mountain. One of the world's most iconic and recognisable landmarks, this flat-topped massif of ancient, weathered sandstone is the first point of land visible to seafarers approaching the Cape of Good Hope, historic gateway to the east.

While suitably inspired by Table Mountain and lacking none of Burchell's anticipatory excitement, our first impressions, 160 years later, were that our new surroundings looked disconcertingly like parts of Scotland. Pounded almost incessantly by cold Atlantic breakers and buffeted by gales that whipped the golden beaches and sheared the heathy moorlands, this rocky corner of the African continent could easily have passed for the Hebrides or some other windswept tract of the country that we had recently and, in the grey chill of winter, gratefully, abandoned.

Our first billet was a field station in the Cape of Good Hope Nature Reserve, not far from the tip of the Cape Peninsula. A few consecutive days of sunshine at this beautiful spot, together with the occasional close encounter with an ostrich and the sometimes over-friendly baboons, reassured us that any similarity to the drizzly north was merely superficial. Here was an environment quite different to Scotland and, indeed, any other.

This difference was manifested most strikingly in the abundance and beauty of its wild flowers. We're not specialist botanists – one of us is an artist and the other an ornithologist, but it didn't take a particularly skilful eye to recognise that, despite much of it looking like a blasted heath, the Cape of Good Hope and the wider landscape beyond was an extraordinary place for plants.

Amongst the bewildering variety of unfamiliar species were a few that even we recognised. Pink pelargoniums, which in our youthful botanical ignorance we knew as 'geraniums', grew wild like weeds, while elegant gladioli and bold blue agapanthus popped up amongst ubiquitous reeds, heaths and scrubby bushes. A suspicion that these had been surreptitiously planted by guerrilla gardeners was soon dispelled by the realisation that the Cape is the original home of a multitude of cultivated flowers that today are denizens of garden and greenhouse.

Having planned to stay at the Cape for one year, 12 flashed blissfully by before we finally prised ourselves away and migrated northwards with the swallows. Helping to soften the blow of being back in Britain was the discovery of nurseries and gardens full of familiar Cape flowers. The ancestors of these had been collected by a succession of plant hunters over three centuries and in whose footsteps we had waveringly followed.

Often as colourful and interesting as their plants, some of the botanical explorers also left detailed and fascinating accounts of their time at the Cape, their journals providing valuable historical insight into a unique landscape and its original and colonial occupants. Many of them also represented the great institutions of Europe, including the Royal Botanic Gardens, Kew, whose first dedicated international plant collector, Francis Masson, was dispatched to the Cape in 1772. The plants that he and others sent back became the subjects of intense scientific scrutiny by academics and, perhaps even more enthusiastically, experimentation and development by the rapidly growing horticultural industry. The profusion of varieties that were bred and marketed reflected the popularity of Cape plants and the skill and industry of the plantsmen of the day.

The diversity of unusual species shipped north in the eighteenth and nineteenth centuries also alerted scientists to the fact that an extraordinary flora lay at the Cape. Here was a botanical treasurehouse ablaze with thousands of species, many of them extremely rare and found nowhere else, whose richness and diversity exceeded all other temperate and the majority of tropical regions.

The Smallest Kingdom is our celebration of Cape plants and the people who discovered and developed many of them into today's horticultural gems. It is by necessity, but unapologetically, an Anglocentric (or Scottishcentric if there was such a word) attempt to give recognition and prominence to the Cape's fabulous flora and its outstanding contribution to the gardens of the world.

ACKNOWLEDGEMENTS

Our time at the Cape was made possible in the first instance thanks to the good offices of Robert Prŷs-Jones, formerly of the Percy Fitzpatrick Institute of Ornithology at the University of Cape Town (UCT) where MF undertook post-graduate research supported by a J. W. Jagger Overseas Student Scholarship.

Derek Clark set us off on the right track. With patience and good humour he introduced us to the many splendours, particularly botanical, of our study site at the Cape of Good Hope Nature Reserve where he was Senior Ranger. His enthusiasm was infectious, and soon infected us.

Friends and colleagues at UCT and other institutions, and members of the Cape Bird Club and the Botanical Society of South Africa, made us welcome and introduced us to the local wildlife over the years. For which, our thanks go to Nigel and Claudia Adams, Alice Ashwell, Mark Beeston, Aldo Berruti Callan Cohen, John Cooper, Richard and Sue Dean, Diana and John Deeks, Atherton de Villiers, Greg Forsyth, John Graham, Tony and Jen Grogan, Phil Hockey, Jenny Jarvis, Richard Knight, Howard and Toni Langley, Belle Leon, Dave Macdonald, Rob

opposite
Foliage, Flowers, and Fruit of the South African Silver Tree
Painting by Marianne North. 'On the right below is a head of female flowers, and above on the left a ripe cone ... in the middle and above are heads of male flowers.'
Official Guide to the North Gallery.

Short-horned grasshopper

Martin, Warren McLeland, Pat Morant, Peter and Miriam Neatherway, Ian Newton, Terry and Margaret Oatley, Mike Picker, Fiona Powrie, Tony and Pat Rebelo, Dave Richardson, Barrie Rose, Walther Seiler, Claire Spottiswoode, Peter and Jenny Steyn, the late George Underhill, Les and Jane Underhill, Etienne van Blerk, Brian and Jane van Wilgen, Barry Watkins, the Watt Pringle family, Gary Williams and Susie Burke, Gerald and Helen Wright, Mark Wright, and John Yeld.

For their friendship and knowledge, generously given, we are especially happy to thank to Richard Cowling and Shirley Pierce, Bill Liltved, and Peter and Coleen Ryan.

Wild flowers for painting were picked under permit from the Cape Department of Nature and Environmental Conservation or were kindly provided from cultivation (notably by Kirstenbosch National Botanical Garden) or private property by John Duckitt, Anthony Hitchcock, Deon Kotze, Howard Langley, Bill Liltved, Neil and Neva Macgregor, Fiona Powrie, Careëne Roux, and Peter Salter.

The gestation of *The Smallest Kingdom* has for the most part been confined to Scotland. Here we have received tremendous help from people at home and, through the wonders of modern technology, abroad. Reference material and other information was helpfully suggested or provided by Richard Cowling (Nelson Mandela Metropolitan University, Port Elizabeth), Matt Daws (Millennium Seed Bank, Kew), Anne-Lise Fourie (Mary Gunn Library, Pretoria), Graham Hardy (Royal Botanic Garden, Edinburgh), Caroline Gelderblom (Working for Water Programme, Cape Town), Roger Hipkin (Penicuik Community Development Trust), and Brian van Wilgen (Council for Scientific and Industrial Research, Stellenbosch). Marcel le Roux and Gill Wheeler helped with translation from Afrikaans.

We thank the photographers credited under their individual pictures, with particular thanks to Richard Cowling, Graham Duncan, Anthony Hitchcock, Steve Johnson, Bill Liltved, Peter Ryan, and Brian van Wilgen who generously allowed us to use their photographs. Ken Paterson expertly helped with digital imaging.

The following helped to track down and supply images, although regrettably we have not been able to include all of them in the book: Ian Apted (University of Tasmania), Helen Bainbridge (Fir Trees Nursery), Carina Beyer and Marcia Marais (William Fehr Collection, Iziko Museums of Cape Town), Helen Broderick and Michele Losse (Kew Library), Jenny de Bruin (Stedelijk Museum De Lakenhal, Leiden), Lydia de Waal and Ulrich Wolff (University of Stellenbosch), David Davidson, Lynne Dibley (Dibley's Nursery), John Duckitt, Joyce Edwards (Amsterdams Historisch Museum), Sylvia Ford (Natural History Museum), Melanie Geustyn (National Library of South Africa, Cape Town), John Graham, Kari Haenow (South African National Biodiversity Institute, Kirstenbosch), Anna Harrison and Hilary Clothier (National Trust Photo Library), Andrea Hart (Natural History Museum), Mats Hjertson (Museum of Evolution, Uppsala University), Lila Komnick and Rayda Becker (Parliament of South Africa, Cape Town), Dianne le Roux (Darling Tourism), Erik Löffler and Elly Klück (Netherlands Institute for Art History, The Hague), Cees Lut (National Herbarium of the Netherlands, Leiden University), Vanessa Miller and Jenny

opposite
Disa 'Marguerite Kottler', a cross between *Disa* 'Hanna Meyer' and *Disa* 'Betty's Bay' and one of an increasing number of orchid cultivars descended from *Disa uniflora* of the western Cape mountains.

Cape dwarf chamaeleons were once quite common in suburban gardens but have been much reduced by pesticides and cats. In the wild, they occur mainly in wooded kloofs, occasionally straying into nearby fynbos.

Auton (Royal Horticultural Society), Elaine Milsom and Andy Neal (Badminton Archives), Syd Neville (The Bowes Museum), Venita Paul (Wellcome Images), Nick Pentz and Cecile Basson (Groote Post Vineyards), Sandra Powlette (British Library), Zaitoon Rabaney and Simone van Royen (Botanical Society of South Africa), Alexandra Read (Penguin Group), Celine Rebiere-Ple (Musée de Louvre), Rod and Rachel Saunders (Silverhill Seeds), Karel Schoemann (National Library of South Africa), Ben Sherwood (The Linnean Society of London), Anna Sjögren (Uppsala University), Mark Spencer (Natural History Museum), Chris Sutherns (Victoria & Albert Museum), Dalila Wallé (Museum Boerhaave, Leiden), Mark Wright and Ken Leonhardt (University of Hawai'i), and John Yeld (*Cape Argus*). Colin Paterson-Jones was especially helpful in providing historical and contemporary images from his collection.

Special thanks to Julia Buckley (Kew Library), Louisa Dare (Courtauld Institute), and John Hunnex (Natural History Museum) for their enthusiastic help with our research. And extra special thanks to Marian Cordry (CBS, Santa Monica) for boldly going where no researcher has gone before to track down an image from Star Trek™. She's a star.

Cultivars of Cape descent for painting were obtained by generosity, stealth or, when all else failed, purchase. We thank Angela Lady Buchan-Hepburn, David Hickson, Stella Findlay, Elizabeth Fraser, and the late Colin Maule for choice specimens.

Gina Fullerlove, Head of Kew Publishing, and her team of John Harris, Lloyd Kirton, Paul Little, Michelle Payne, and Lydia White, together with designer Lyn Davies and editorial consultant Alison Rix, skilfully steered the book through to publication and have been a pleasure to work with.

A number of original paintings from *The Smallest Kingdom* have been purchased by The Shirley Sherwood Collection, the proceeds subsidising the costs of publication. We thank Dr Sherwood for her interest in and support for the project.

For better or worse, our friends and family members didn't see or hear a great deal of us while we were working on this book. For their forbearance and much appreciated encouragement and moral support we are indebted to them all and, in particular, to Steve Bales, the late Ian Balfour Paul, Nancy Gordon, Ken Paterson, Gill Wheeler and, not least, our mums Joan Fraser and Janet McMahon. And a very special mention in dispatches for Lesley Shackleton.

Plant names

In most northern hemisphere countries virtually every plant has the luxury of at least one name in the local language. In contrast, very few Cape plants have English or Afrikaans names, and many of these often refer to more than one species. The use of Latin names in the text is not, therefore, scientific ostentation but simply because they are the only ones available for the great majority of Cape plants.

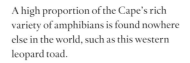

A high proportion of the Cape's rich variety of amphibians is found nowhere else in the world, such as this western leopard toad.

Lambert's Bay

Elands Bay

◆ Clanwilliam

Cederberg

Kouebokkeveld Mountains

GREAT KAROO

Cape Columbine

St Helena Bay

◆ Piketberg

Saldanha Bay

Langebaan

Grootwinterhoek Mountains

◆ Karoopoort

Witberge

Elandsberg

Tulbagh ◆

◆ Ceres

Riebeek-Kasteel ◆

Klein Swartberg

Darling ◆

SWARTLAND

Hex River Mountains

Dassen Island

◆ Malmesbury

◆ Paardeberg

Bokbaai ◆

◆ Worcester

◆ Paarl

Robben Island

◆ Franschhoek

◆ Robertson

Langeberg

LITTLE KAROO

Cape Town

Table Bay

◆ McGregor

Langeberg

Table Mountain

Cape Flats

◆ Stellenbosch

Hout Bay ◆

◆ Greyton

◆ Swellendam

False Bay

Hottentots Holland

◆ Riviersonderend

CAPE PENINSULA

Simon's Town ◆

◆ Caledon

Cape of Good Hope Cape Point

OVERBERG

◆ Riversdale

Cape Hangklip

Cape Infanta

◆ Bredasdorp

Danger Point

◆ Arniston

◆ Elim

Quoin Point

◆ Struisbaai

Cape Agulhas

NASA/courtesy of nasaimages.org

pernâ magis, ut quod explicatum, ex multis squamis imbricum modo invicem sibi incumbentibus constans, quæ circa mediam & supernam ipsius capitis partem majores, præ siccitate reflexæ erant summo fastigio, & quodammodo in lacinias divisæ, coloris spadicei: inter quas exortæ erant veluti oblongæ membranaceæ ligulæ, planæ, & satis angustæ, cineracei quasi coloris, summâ parte latiores & foris exalbidæ, internè autem spadicei coloris, mollissimo longoque villo saturæ purpuræ elegantissimæ per oras ornatæ, quæ temporis successu in atrum degenerabat. In medio capite densa erat villosorum staminum flavescentis coloris congeries, inter quæ latebant angustæ & pungentes spinæ, binas uncias longæ, quibus infimâ parte villosum semen inhærebat, Nerij semini non dissimile, fusci ex rufo coloris, sed immaturum. Hunc cum XVII. & XVIII. in eadem tabella, quæ est Tertia hujus capitis, expressum proponimus.

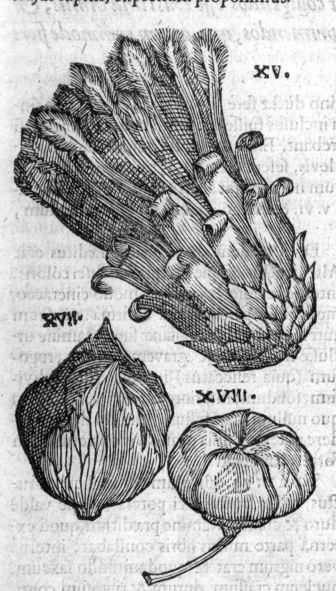

XV.

XVII.

XVIII.

XVI.

Tianche

Fructus XVII

Fructus XVIII.

SED & flosculi in illis capsis erant, qui an in arbore nati essent, an in aliqua alia planta, me latet: sed charta, in qua involuti erant, *Tianche* inscriptionem habebat: candidúsne verò an flavus fuerit flos, dubium: nam ex resiccato difficilis erat dignotio, præsertim tam brevi momento pereunte nativo colore: calicem autem habebat è quatuor veluti squamis compactum, & brevi petiolo præditum, in quo flos erat quatuor foliis longioribus & calicem superantibus constans, angustioribúsque quàm calicis squamæ, decussatim compositis, protuberante in medio umbone, ut in malorum aureorum & citriorum floribus: odor (si quem habuerat) evanidus, quemadmodum & sapor, propter siccitatem: ab incolis tamen in aliquem usum adservatos fuisse, verisimile est. Eos porrò Decimum-sextum locum obtinere voluimus.

ADMIRABILIS formæ erat Decimus-septimus: quatuor enim membranaceis foliis ex adverso inter se oppositis, ex eodem tamen pediculo prodeuntibus, constabat, quæ supernâ & extimâ parte veluti coëuntia, osficulum, infimâ parte ipsi inhærens & fragile occultabant, in quo nucleus orbicularis formæ nucis moschatæ magnitudine continebatur, foris fuscâ & fungosâ quadam materiâ obsitus, durus, ut qui resiccatus, internè albus & corneæ duritiei: membranacea illa folia, multis venis erant distincta, & rhomboidis formæ, atque uncialem longitudinem & latitudinem habebant. Nullum ejus nomen indicare poterant Indi ab Hollandis advecti, quemadmodum multarum aliarum rerum: sed quæcumque ignorabant (ignorabant autem plurima, ut qui omnes vilia mancipia fuissent, præter Abdalam, quem in Iava mercede conduxerant) interrogati quî vocarentur, *Madagascar*, respondebant, tamquam innuere volentes, apud ipsos non nata, sed è Madagascar, sive D. Laurentij insula, esse delata, quia naucleros Hollandos in ea aliquot dies hæsisse non ignorabant; imò illorum natu minimus inde advectus fuerat.

SVBROTVNDÆ in plano formæ erat *Decimus-octavus* fructus, uncialis ferè, nigro aut subfusco & lævi cortice tectus, in quinque partes diviso, sub quo latebat cartilaginosus alius cortex, inæqualis & asper, extuberans eâ parte quâ superior cortex divisus, continens

CHAPTER ONE

The Thistle from Madagascar

*At every step a different plant appeared;
and it is not an exaggerated description if it should be compared
to a botanical garden, neglected to grow in a state of Nature;
so great was the variety everywhere.*

The English naturalist and traveller, William Burchell, wrote these words in his journal after a short walk along the western slopes of Table Mountain on a hot afternoon in November, 1810.

At this time of year many of the plants would not even have been in leaf, let alone flower, but he still recorded over 100 different species in less than a mile. 'I believe that double that number may be found on the same ground by searching at different times', he added.

Burchell was not the first visitor to enthuse about the Cape countryside, but it took almost 300 years from the first European discovery in the late 1400s for the region's unique botanical qualities to be fully recognised and appreciated by the outside world. Nevertheless, the discovery and exploration of the Cape are, one way or another, intimately entwined with its remarkable plants.

For a long time Africa was considered little more than an annoying obstacle to trade and the exploration of the east. In 1844, the historian Hugh Murray declared that:

Could Africa cease to exist, great facilities would be afforded to communications between the other continents, and many new channels of commerce be opened up.

Being obstinate, as major landmasses are wont to be, Africa stayed put, so anyone curious enough to know what lay beyond had no option but to sail south and round the Cape of Good Hope.

Today, the Cape of Good Hope refers specifically to the most southwesterly of the three sandstone headlands where the Cape Peninsula dips its gritty toes into the cold Atlantic. To the early navigators, however, it was the flat-topped massif of Table Mountain and, beyond it to the south, the Cape Peninsula separating the open ocean to the west from the broad inlet of False Bay to the east.

Conventional history records that the Cape of Good Hope was first discovered by Europeans in 1488. By this time the Portuguese, under Prince Henry the Navigator and King John II, had already set in motion a succession of pioneering journeys down the west African coast. In addition to general colonial expansion, their aims were to find a passage to the east to outflank the Arab trading routes of the Middle East, and to loosen the stranglehold that Venice and Genoa had on the valuable commodities – the spices, tea, silks, minerals and ceramics – that came overland from India and the Far East.

The great Portuguese seafarer Diago Cão had pushed the limits of exploration south to Cape Verde by 1445, the mouth of the Congo (Zaïre) River by 1484, and Namibia by 1486. Late in the following year, Bartolomeu Dias struck even further south and became the first European to journey to the southern edge of Africa.

Dias was, at first, unaware of his achievement. Having made brief landfall just beyond the Orange River (which today forms the border between South Africa and Namibia), his small fleet then beat into a southerly gale with sails at half-mast. Making a wide tack out to sea, they sailed out of sight of land for 13 days. When the wind eased Dias turned east, expecting to hit the coast once again. Finding nothing, after a few days he headed north and soon encountered the southern Cape coast near Mossel Bay, some 350 km east of Cape Town. Seeing that the lie of the land had swung from north-south to west-east, Dias realised that he must have turned the corner of the continent.

---

opposite

'The Thistle from Madagascar'
A dried flowerhead of Blue Sugarbush *Protea neriifolia* ('XV') was the first Cape plant to be brought to Europe. Collected by Dutch seafarers on the southern Cape coast in 1597, the specimen is featured in the *Exoticorum*

Libri Decem ('Ten Books of Exotica') by Charles de l'Escluse (1529–1609). This species does not occur in Madagascar, but Dutch ships did call at the island on their trading voyages to Java, so the mistaken provenance is understandable.

On the return journey from the eastern Cape, Dias erected a limestone cross or *padrão*, a symbol of ownership and a point of reference, somewhere on the Cape Peninsula on Saint Philip's Day, 6 June 1488. This *padrão* has never been found but the remains of those that Dias set up at Kwaaihoek at the eastern limit of his voyage on 12 March and at Dias Point near Luderitz on 24 July 1488 were discovered by historian Eric Axelson in 1938 and 1953, respectively. *Cape Archives*

Dias's fleet consisted of a supply ship and two caravels, the latter small vessels of only 45 tonnes displacement. They were not designed or equipped for the open ocean, but lateen-rigging (small, triangular sails set at 45° to the mast) allowed them to tack, and a shallow-draft enabled close approach to the coast to investigate bays and inlets. It had already been a long journey under difficult conditions and by the time the fleet got as far east as Cape Infanta (which he named after João Infante, captain of one of the ships) and briefly put ashore, Dias's crews had had enough and he was persuaded to turn about.

Early on the return leg of the voyage, probably in April 1488, the fleet at last caught sight of the cape at the south-western tip of Africa. Historians are divided as to whether it was Dias or, later, King John who christened this important and distinctive landmark the 'Cape of Good Hope'. Some say that Dias called it, with justification, *Cabo Tormentoso* – the Cape of Storms; others prefer to believe that it was King John who renamed it Good Hope 'because it gave promise of the discovery of India, so long desired and sought for many years.' Either way, Dias's chronicler, João de Barros, relates how the expedition finally 'came in sight of that great and famous cape concealed for so many centuries.'

Without prior knowledge, one wonders how the Portuguese could have had any preconception of such a 'great and famous cape' at the extremity of Africa? Henry the Navigator and King John II had certainly appreciated that a new world and a sea route to India would open up once a *Promontorum prassum* that lay at the tip of 'Libya' (as the African continent was sometimes known in those days) was attained. '*Promontorum*' is just what it sounds like – a promontory; '*prassum*' is seaweed or, more specifically, the kelp which grows in the coastal waters of the Peninsula. In a nutshell, 'a promontory surrounded by kelp' would perfectly describe the Cape Peninsula and the Cape of Good Hope and, in fact, not too many other spots on the South African coastline. This would imply that the Portuguese somehow already knew of such a place, despite there being no historical documentation of previous visitors to the Cape from Europe.

Such knowledge may, however, have come to them from Indian or Chinese explorers, as it is very likely that they had made earlier visits to the Cape. A map that pre-dates Dias's visit by almost 50 years has the name 'Cabo di Diab' (Devil's Cape) at the tip of a crude but clearly recognisable rendering of the African continent. Some historians have

---

On Wednesday the 11th [March 1620] we saw 'Trombes', which are large seaweeds about three or four fathoms long, also birds that the Portuguese call 'alcatras' or 'margauts', which have white bodies and only the tips of the wings black, cormorants, sea-bears, and penguins, any one of which is a sure sign that land is not far off.

… I took a walk behind Table Mountain, and went about three leagues into the country. I noticed that the soil was very good, and saw that a little stream of fresh water wanders thought the open plain and enters the seat the head of [False Bay] … Having walked until midday day in this country, which is covered with grass and lovely flowers, I took my way along the mountains and came to the foot of

the Table where it faces south. There I found a great many trees, including some for which planks a foot wide and eighteen to twenty feet long could be cut.

… All along the mountains there is an infinity of game, such as roebucks, deer as large as harts, partridges, and all sorts of game, and on the mountains are great numbers of monkeys, marmots, lions, lynxes, porcupines, ostriches, elephants, and other beasts unknown to me.

Augustin de Beaulieu, who sailed from Honfleur in a fleet comprising *Montmorency*, *Espérance* and *Hermitage*, on 2 October 1619.

pushed this appellation as far back as 1306, and the Greek historian Herodotus describes how Phoenician sailors, commissioned by the Egyptian pharaoh Necho, claimed to have circumnavigated the African continent in a fleet of galleys in the sixth century BC.

More prosaically, it may have been that the Portuguese had confused their coasts and were actually referring to the 'Cape Prassum 'in Moçambique. Now known as Cape Delgado, this location featured on the maps of the astronomer and geographer Ptolomey of Alexandria and was, at the time (the second century AD), well frequented by Arab traders but the most southerly point of Africa known to Europeans.

Fat-tailed Sheep and a Basket of Bulbs

Dias arrived back in Lisbon in December 1488 but it was almost ten years before the Cape received a second recorded visit from the Portuguese. There may, however, have been other expeditions that were kept under wraps for commercial reasons.

Vasco da Gama set sail on 8 July 1497 with a fleet of four vessels, including a supply ship. Dias himself accompanied the expedition as far as the Cape Verde Islands. The other three ships came within sight of the Cape of Good Hope on 18 November and rounded it four days later. Da Gama thereafter continued his voyage, sailing up the east coast of Africa as far as the Kenyan city of Malindi where he found an Arab pilot who guided him the 3,000 km across the ocean to India, arriving at Calicut in May 1498.

Although the Cape was a halfway house between Europe and the Far East, very few vessels appeared to call there during the 100 years following Dias's and da Gama's voyages. The few surviving written records, such as logs and personal diaries, describe how passing ships made brief stops at Saldhana Bay (just over 100 km north of Cape Town), Table Bay or Mossel Bay to replenish water. A general unwillingless to linger at any of these spots resulted from the unpredictable but frequently stormy weather, the treacherous currents, and an absence of valuable minerals or spices to buy from the Khoekhoen, the Cape's aboriginal inhabitants.

The Khoekhoen were originally christened 'Hottentots' by the Europeans in the seventeenth century, apparently because of their unusual 'click' language. The early travellers also knew them as the Khoikhoi, and this form of their name persisted until relatively recently.

Seasonally-nomadic herders of the Cape coastal plains, the Khoekhoen kept large numbers of cattle and fat-tailed sheep, bartering these for metal items, cloth and trinkets. Relations between the Europeans and Khoekhoen were often strained by difficulties in communication and, not least, the generally superior attitude adopted by the visitors.

Early in the history of dealings between the two cultures an unhealthy standard was set by Francisco d' Almeida, first Viceroy of Portuguese India. In 1510 his fleet of 13 vessels stopped for freshwater and livestock on the coast not far north of the Cape. After an apparent misunderstanding over the price of a cow, or following some pilfering (the reasons differ between reports), d'Almeida went ashore with a punitive force but was killed, along with 50 of his men, in the ensuing skirmish.

D'Almeida was buried on the shores of Table Bay and two years

… he climbed a mountain, very flat and level on the top, which we now call 'The Table of the Cape of Good Hope', from whence we saw the end of the Cape, and the sea that lies beyond it to the east, where there was a very large bay, where, at the end of it between two ranges of high rocks which today we call 'Os Picos Fragasos', he saw a large river which seemed to have a long course, since it had much water, and from these landmarks he knew that it was indeed the Cape of Good Hope, so that with the first favourable wind they doubled it, now going on their way with more confidence.

João de Barros (1496–1570) describing the voyage of Vasco da Gama, 1495–97

De Brito marks d'Almeida's Grave
This retrospective representation of an event that occurred in 1512 was published by Pieter van der Aa in 1707. It is fanciful, to say the least, but the idea of someone collecting plants (bottom right) must have been based on more than artistic whimsy.

later his grave was marked by a large wooden cross and a cairn. This event was depicted in an engraving, 'De Brito marks d'Almeida's grave', published by Pieter van der Aa in 1707. Although the illustration is largely fanciful and depicts little of historical or geographical accuracy, it is at least interesting to note that a minor character in the bottom right of the tableau is gathering flowers and, perhaps, bulbs, and putting them in a basket. It could be that these were to adorn the grave or were for eating, but it is just as likely that plant collecting was recognised as a regular activity at the Cape by 1707, if not in 1510.

The Fairest Cape

A number of factors, including the d'Almeida incident, dissuaded the Portuguese from establishing a permanent station at the Cape and they switched their attention to the east coast of Africa, setting up refreshment facilities there. By the end of the sixteenth century, however, and as other seafaring powers began to make their move, they had lost the initiative on the Cape route.

On the homeward leg of his epic global circumnavigation, Francis Drake rounded the Cape of Good Hope on 18 June, 1580, an event recorded as follows:

> We ranne hard aboord the cape, finding the report of the Portugals to be most false, who afferme that it is the most dangerous Cape of the world, never without intolerable storms and present dangers to travellers which come neare the same. The cape is the most stately thing and the fairest cape we sawe in the whole circumference of the earth.

'The Fairest Cape' has become the favourite publicity slogan of those whose job it is to entice visitors there, this despite the facts that, contrary to popular belief, it wasn't Drake who actually coined these words but the expedition chaplain, Francis Fletcher, and that those aboard *Golden Hind* must have been blessed with an uncharacteristically calm day.

The English push to the Far East was given extra impetus by Drake's successful voyage and emboldened by the defeat of the Spanish Armada in 1588. Three years later the first English fleet hove to off the Cape and in due course their vessels became more regular visitors as they made inroads into the lucrative trade of India and the Far East.

The Thistle from Madagascar

In July 1595, four Dutch ships under the command of Cornelis de Houtman made the first commercial trip from Holland to the East Indies. The fleet returned in 1597 and a report of the expedition was published the following year.

De Houtman's fleet stopped at Mossel Bay on 6 August 1595 to take on fresh water. Here, a member of the ships' company 'was sent to examine the land, which indeed offers a fine prospect, adorned with sweet-smelling shrubs and flowers.' For its time, this was an almost unique reference to the local plantlife, and additionally unusual in being

an emotional response rather than the more traditional and practical allusion to 'inedible herbage' or the like. Most significantly, it is likely that this same spot was the source of the first Cape plant to reach Europe.

Collected in 1597 by an unknown sailor from an equally anonymous Dutch merchant ship, this pioneering plant eventually made its way to Professor Charles de l'Escluse, usually known as Clusius, director of the botanic garden at the University of Leiden. His reaction on seeing the specimen is not recorded, but it clearly made enough impression to be featured in his 1605 publication *Exoticorum libri decem, quibus animalium, plantarum, aromatum, aliorumque peregrinorum fructuum historiae describuntur*. A catchy title, typical of the age, this translates as 'Ten books of exotica: the history and uses of animals, plants, aromatics and other natural products from distant lands.' From the illustration, the plant in question can be identified as a dried flowerhead of *Protea neriifolia*, a shrub that grows in dense stands along the southern Cape coast.

Proteas are found only in sub-Saharan Africa and such a specimen would have been completely alien to Clusius. The best he could do was to describe it in terms of a plant with which he was familiar, or to which he thought it might be related. So, Figure XV in *Exoticorum Libri Decem* became '...the head of a certain very graceful genus of *Carduus* [thistle]...collected at...the Isle of Madagascar'. The dried flowerhead is, I suppose, roughly the shape of an oversized thistle. There are no proteas in Madagascar, but the error of provenance can be forgiven because that island was a place:

> where the Hollanders on their first sea-journeys procured water and other necessary things for their food and where they were staying a few days before they left for Java.

Its confused origin and identity notwithstanding, this 'graceful thistle' was, as far as we know, the first representative of the Cape flora to reach Europe. Alive or dead, proteas are famously tough and leathery, characteristics that would have allowed the specimen to travel intact so far and for so long. As a member of the quintessential Cape family – charismatic, beautiful and resilient – there could not have been a more appropriate ambassador of the botanical treasure house that awaited discovery.

De Houtman's expedition was commercially unprofitable and a humanitarian disaster; only 87 of the 240 crew survived, and much suffering and bloodshed was inflicted upon the indigenous inhabitants of the places they visited. A navigable route to India and the islands of the Far East had, however, been established and the potential was recognised for enormous profits to be made from ships that successfully made the voyage and returned laden with valuable cargoes.

This was 'third time lucky' for the Dutch. Embroiled in a protracted war of independence with the Spanish, they were short of money and desperately needed to find a way to the Far East to establish their own business and boost a flagging economy. This necessity was compounded by the Spanish capture of Portugal and their closing of Lisbon – the main spice-trading harbour in Europe – to the Dutch. In 1596 a Dutch fleet had attempted to find a route to the East round the north of Russia, only to become trapped in the polar ice for the winter. The many fatalities included Captain Barents, after whom the sea is named. In an attempt to find a route around the tip of South America through the Straits of Magellan a five-strong fleet left Rotterdam in 1598. One ship got lost and returned without having got anywhere; one vanished in the Pacific; one surrendered to the Spanish, while another was captured and its crew put to death; and one arrived in Japan with 24 of its 110 crew alive. The Cape route was, therefore, a godsend and the Dutch merchants were delighted. The nautical bandwagon, already well under way (15 fleets amounting to 65 ships sailed from Holland to the East Indies between 1595 and 1601), soon gained pace.

On a second voyage to the East in 1598, de Houtman had as his chief pilot the British seafarer John Davys. In his account of the voyage, Davys observed on 11 November, at what was later to be known as Table Bay (now Cape Town's harbour), that:

> This land is a good soile, and an wholesome Aire, full of good herbes as Mintes, Calamint, Plantine, Ribworte, Trifolium, Scabious and such like.

It is not clear what these plants are, but presumably they resembled species with which Davys was familiar back in England and all of them would be well known to the apothecary or herbalist. Mint, for example, was a popular antiseptic and infant cordial and also used to treat hiccups and haemorroids, while scabious was the favoured remedy for pleurisy and dandruff.

Most importantly, Davys' plants could also have been useful in the treatment of scurvy, the scourge of seafarers whose diet lacked Vitamin C. This occurs in particularly high concentrations in citrus fruits, tomatoes and potatoes, but is found in lesser quantities in most fresh fruit, roots, and leaves. Plants found at the Cape that had any medicinal or nutritional use would thus have been of the greatest, or only, interest to sailors weakened and wearied by months at sea. The multitude of unusual and unfamiliar Cape plants confronting anyone who stepped ashore would simply have been ignored or unrecorded if they did not bear some resemblance to something therapeutic or vaguely edible.

The First Cape Collection

While any interest in the plant life of the Cape was likely still to be restricted to its usefulness or otherwise, the development of post-Renaissance botany as a study of plants in their own right continued apace in Europe. Here, physicians were evolving into pure botanists, and physic or apothecaries' gardens into botanical gardens.

The study of plants as scientific curiosities or things of beauty beyond their practical value was initially the indulgence of the wealthy few – royalty, aristocrats and merchants who increasingly benefited from the global expansion of trade and colonialism. They were, in turn, able to patronise the leading botanical experts and the new breed of nurserymen. A broadening interest and soaring economic prosperity, therefore,

Charles de l'Escluse (sometimes spelt 'de l'Ecluse', and known in academic circles as Carolus Clusius), was appointed professor of botany at the University of Leiden in 1539. He established a garden there that he filled with plants from the Americas and, in particular, the Near and Middle East. His numerous introductions, including tulips, hyacinths and naricissus, were developed into a horticultural industry that continues in Holland to the present day. His Cape connection, 'The Thistle from Madagascar', is limited but highly significant in the history of plant collecting.
Wellcome Library, London

combined to nurture the emergence and upsurge of horticulture and an obsession with plant collecting and consequent demand for new and exotic species.

The voyage of De Houtman and Davys accounted for only one of 22 Dutch ships that sailed round the Cape to the Far East in 1598. A further increase in traffic and trade culminated in the establishment, in 1602, of the Dutch East India Company, the VOC (Verenigde Oostindische Compagnie; United East Indian Company). The British had already established their own Honorable East India Company in 1600, but the VOC was of most significance as far as the history of plant collecting at the Cape is concerned.

In his botanical garden at Leiden, Clusius had already built up an unrivalled variety of plants, including many species from the New World, and bulbs such as narcissi, tulips, and hyacinths from Turkey and the Levant that were to be the makings of the highly lucrative Dutch horticultural industry. Because such decorative plants were becoming increasingly sought after and valuable, Clusius was able to use his growing influence to persuade the directors of the VOC to instruct their captains and crews to collect material on their travels whenever and wherever possible.

The great majority of new species that now reached Holland by the Cape route came from the Far East where the Dutch had wasted no time in establishing settlements and trading stations. The first of these was in Banten, western Java in 1601, followed by Jakarta (capital city of the Indonesian archipelago and known as Batavia until 1942), in 1610. A trickle of plants from the halfway house at the tip of Africa, however, grew gradually into a flow.

By 1605, a variety of Cape bulbs could be found blooming in Dutch gardens. Probably the first to do so was a species of chincherinchee *Ornithogalum* grown in Amsterdam and which, on flowering, was brought to Clusius. It later featured in his *Curae posteriores aethiopicum*, published posthumously in Antwerp in 1611 (he died in 1609). Here it is mentioned that the bulbs were gathered 'at that extreme and celebrated Promontory of Aethopia commonly called the Cape of Good Hope'.

While the person who brought back these particular bulbs remains anonymous, a name that has been recorded for posterity is that of Gouarus de Keyser. A botanical treatise by Mathias de l'Obel records how de Keyser dug up some bulbs at the Cape in 1603 and brought them back to his brother, Jacobus, a merchant and keen plantsman living in Wiesbaden. De Keyser thus becomes the first known plant collector at the Cape.

De l'Obel's voluminous compendium contained over 2,000 illustrations, including de Keyser's Cape plants. The latter were named *Narcissus Aphricanus folio rotundiore* ('African daffodil with round leaves') and *Narcissus Aphricanus bifolius* ('African daffodil with two leaves'). Although one illustration has a bulb with no flower, and the flower in the other is inconsistent with the species, the bulbs can be tentatively identified as *Haemanthus sanguinens* and *H. coccineus*, respectively. *Haemanthus* means 'blood flower', and is a genus of lily-like amaryllids

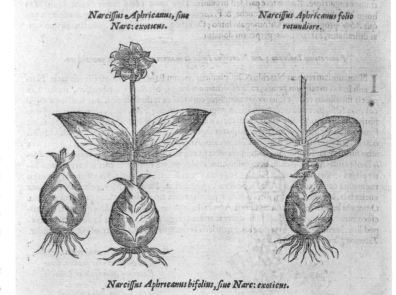

Species of *Haemanthus* featured in a number of early botanical publications, including one in 1605 by Mathias de l'Obel (1538–1616), after whom *Lobelia* is named. Although quite crude, these drawings can be identified as *H. coccineus* (left) and *H. sanguineus*. If there was any doubt as to the plants' colour these names translate as, respectively, 'Deep-red blood-flower' and 'Blood-red blood-flower'.

The trademark of the Dutch East India Company, the Verenigde Oostindische Compagnie.

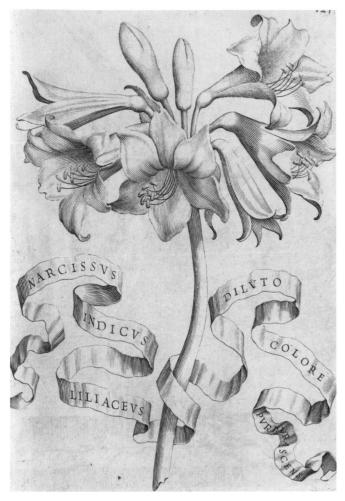

Giovanni Battista Ferrari (1584–1655) was in charge of Barberini Palace garden in Rome which boasted one of the finest collections of rare species arriving from newly explored parts of the world. In 1632, Ferrari published *De Florum Cultura*, essentially a gardening book that represented an important step towards growing plants for pleasure and beauty rather than herbal medicine. Four Cape species are depicted: *Amaryllis belladona* (left), *Brunsvigia orientalis*, *Haemanthus coccineus*, and *Ferraria crispa* (right). All of these would have been considered remarkable in Europe at the time, particularly the *Ferraria* with its curiously crinkly-edged, purple-and-yellow flowers and strong scent of vanilla. Later introduced as a garden ornamental to Australia, it has become so successful there that it has been declared a noxious weed.

characterised by a large bulb and usually two broad leaves up to 70 cm long. De l'Obel recorded that he:

> saw this plant of rarest elegance among the exotic delights of the gardener I. Knibius of Middleburg, Zealand [part of the Netherlands] but it perished after the severe winter.

Although there are rather few detailed records of what was being brought back in the early years of the seventeenth century, it is apparent that bulbs (in which the Cape is extraordinarily rich) soon became much sought after in Holland and elsewhere in Europe. Enough were certainly being imported and propagated to be distributed among the increasing number of gardening enthusiasts and their patrons.

It also appears that any horticultural rivalry between the various countries did not prevent even the most rare and prized bulbs from being freely exchanged and distributed. A result of this was that Cape plants appeared not only in gardens but in a variety of botanical and horticultural publications.

As illustrations were frequently copied from one publication into another, it is possible to trace the route of some plants, differing only by a leaf here or a petal there, through the years and over national boundaries into diverse works of botanical scholarship. One of de Keyser's *Haemanthus*, for example, having featured in de l'Obel's work in 1605, reappeared in Giovanni Ferrari's *De florum cultura libri*, an Italian publication of 1633. Ferrari, a Jesuit priest, botanist and linguist, also included detailed and accurate illustrations of belladonna lily *Amaryllis belladonna*, candelabra flower *Brunsvigia orientalis*, and spider iris *Ferraria crispa*.

Considered to be the first commercial horticultural catalogue, the *Florilegium* of Emmanuel Sweert (1552–1612) advertised his plants at the Frankfurt Fair of 1612 and includes three from the Cape. The intensity of colour of the illustrations varies between volumes, the 'Gladiolus maximus', for example, ranging from monochrome to deep red. It is not possible to identify it to species, although *Gladiolus carneus* seems most likely.

The last two are particularly unusual and must have made a lasting impression on anyone who saw them.

Commercial plant catalogues, notably that of Emmanuel Sweert, first appeared as early as 1612. A Dutch nurserymen, Sweert worked at the Vienna gardens of the wildly eccentric Emperor Rudolf II, who is perhaps better remembered for his collection of dwarfs and regiment of giants. Sweert's *Florilegium*, a trade publication advertising his stock at the 1612 Frankfurt Fair, included three Cape species – *Boophane disticha*, *Drimia capensis* and a gladiolus, possibly *Gladiolus carneus*.

It is said that Sweert destroyed bulbs in order to maximise their rarity value, as new bulbs were becoming so popular that they could command increasingly high prices. If there ever was a gentle age of gardening generosity it was now, by all accounts, giving way to a hard, business-like approach.

Delighteth in the Moones Appearance

Whether through the remaining vestiges of horticultural philanthropy or through expanding trade, the early years of the seventeenth century saw Cape plants spreading across Europe. In 1633, a revised edition of John Gerard's 1597 classic *The Herball or Generall Historie of Plantes* included the following reference:

> There is of late brought into his kingdome, and to our knowledge by the industry of Mr John Tradescant, another more rare and no less beautifull than any of the former, he had it by the name *Geranium indicum noctu odoratum* … At the top of each branch upon foot stalkes some inch long grows some eleven or twelve floures, and each of the floures consisteth of five round pointed leaves of a yellowish colour with a large blacke purple spot in the middle of each leafe as it were painted, which gives the floure a great deal of beauty; and it also hath a good smell. I did see it in floure about the end of July 1632 being the first time that it floured with the owner thereof. We may fitly call it the Sweet Indian Storksbill or painted Storksbill; and in Latin *Geranium indicum odoratum flore maculato*.

This is clearly *Pelargonium triste* from the Cape.

John Tradescant the Elder, pioneer plantsman, inveterate collector and 'Keeper of his Majesty's Garden, Vines and Silkworms' at Oatlands Palace, Surrey, obtained *Pelargonium triste* seeds from René Morin, head of a distinguished family of French gardeners, in 1631. Tradescant was, therefore, responsible for introducing 'geraniums' to England, even if this particular species did not gain the wide and lasting popularity in cultivation later enjoyed by the many others of its clan, properly known as pelargoniums, from the Cape. The reference to India in Gerard's account is a fairly common error of provenance, given that many plants were arriving from that part of the world at the time, with Cape ones doubtless often mixed in and poorly labelled.

In 1629, John Parkinson's *Paradisi in Sole Paradisus Terrestris* ('Park-in-sun's Earthly Paradise' – an early, and distinctly unmemorable, horticultural pun) was published. This popular work is considered to be the first real gardening book to appear in England, one that viewed

The frontispiece from the 1636 edition of Gerard's *Herball*, first published in 1597 and revised by Thomas Johnson, 'citizen and apothecarye of London', in 1633.

NARCISS. PVMILVS INDIC. POLYANTH.

ORNITHOGAL. LVTEOVIRENS INDIC.

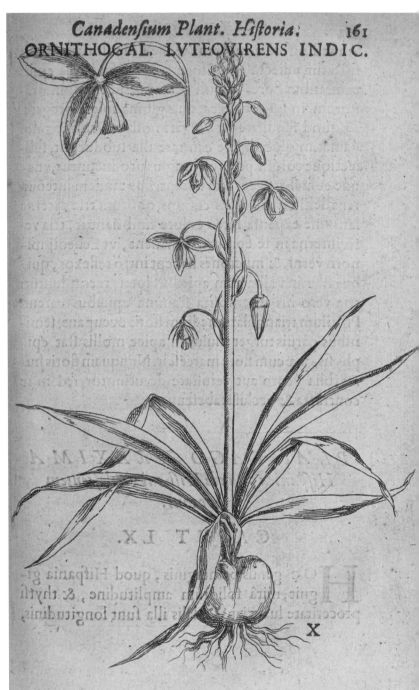

The title of Jacques-Philippe Cornut's book *Canadensium plantarum,* and the names of some of the plants featured in it reflect prevailing confusion over their true origin. Six Cape species are illustrated, including *Chasmanthe aethiopica*, Malgas lily *Ammocharis (Cybistetes) longifolia*, Guernsey lily *Nerine sarniensis*, and *Albuca canadensis* (now *flaccida*). Cornut (1606–1651) was a French physician who obtained plants from the Royal Garden of Henry IV and from Jean Morin, a leading Parisian nurseryman.

and discussed plants primarily from aesthetic and decorative perspectives. Dedicating it to Queen Henrietta Maria earned its author the grand title of *Botanicus Regius Primarius* from her husband, Charles I, but, more importantly from our point of view, it contained a further reference to *Pelargonium triste*. Parkinson observed that:

> The flowers smell very sweete like Muske in the night onely and not at all in the day time as refusing the Sunnes influence but delighteth in the Moones appearance.

Such poetic sentiments convey an enthusiasm and affection for the species that belie its generally unassuming looks and almost weed-like habits and abundance in sandy coastal areas of its native western and southern Cape.

Parkinson also refers to the 'Starre-flower of Aethiopia', being the chincherinchee described by Clusius, and the 'Sea-daffodil of Africa' which is de l'Obel's *Haemanthus*. Both were apparently growing in Parkinson's collection.

Across the English Channel, meanwhile, Cape plants made an appearance in a French book, published in 1635. Jacques-Philippe Cornut's *Canadensium plantarum, aliarumque nondum editarum historia* includes *Chasmanthe aethiopica*, Malgas lily *Cybistetes longifolia*, Guernsey lily *Nerine sarniensis*, *Albuca canadensis* (now *flaccida*), and *Romulea rosea*. The common names of some of the Cape species, for example *canadensis*, Guernsey and Malgas (Madagascar), betray their confused origins.

It is significant to note that the first three of these plants are autumn-flowering. As such they would have been very conspicuous, even to the most untrained or disinterested eye, amongst the otherwise dull and desiccated vegetation at the end of the gorgeously long, hot Cape summers. This is the time of year when the south-eastern trade winds allowed the return journey from the Far East to be made with relative ease and speed, and many of the Dutch and British ships called at the Cape in this season. Spring-flowering bulbs are abundant and spectacular at the Cape, but any collected on the outward voyage would have been unlikely to survive a round trip via the Far East that could have taken a year or more. By autumn, the spring plants would have been dormant and difficult to find other than by accident when digging up something else. Plant phenology and trade winds across the Indian Ocean combined, therefore, to shape the gardens of kings and collectors at this time.

The Earth Abounds with Roots

While Cape plants were receiving attention in Europe, it is perhaps surprising that so few observations were made of them on their home ground. One of the few early reports was by Cornelius Matelief who described how 'On the 20 [of April 1608] the Admiral sent ashore the half of a cask which had been sawn in two, to set therein some flowers and earth.' Having headed inland, the plant-collecting contingent was soon approached by some opportunistic Khoekhoen who 'cleverly snatched the shovel from their hands and ran off with it.'

Unsuccessful as this particular venture may have been, it is evidence

One of the main reasons that particular plants were first gathered at the Cape is that they could be eaten by scurvy-ridden sailors, long deprived of fresh greenery. The lower stem of *Albuca altissima* is edible but very slimy, giving it the Afrikaans name of 'slymstok'.

that plant collecting by the Dutch might, by this time, have become routine and, despite a lack of contemporary written records, quite a lot of material was probably reaching Europe. Only if a plant was later featured in some botanical publication was there ever a hint as to when it was first imported. For example, *Knowltonia vesicatoria*, a green-flowered member of the buttercup family found widely in the Cape, appeared in a manuscript by de l'Obel, incomplete at the time of his death in 1616, that was published under his name by the physician William How, nearly 40 years later. According to How, the plant had come from the London garden of John de Franqueville sometime between 1590 and 1610.

In general, if they were mentioned at all in seafaring records, plants give the impression of being afterthoughts to the major business of navigation and trade. Nevertheless, occasional botanical observations did provide some small insight into the plant life of the Cape that may have aroused the curiosity of botanists back home. Nicholas Downton, captain of the *Pepper Corn*, investigated the base of Table Mountain in 1610 and recorded:

> … it is moist ground, and seemeth to be good pasture for cattle, in divers places scatteringlye we see some trees of small stature, somewhat broad topped, bearing a fruit in bigness and proportion like a pineapple, but the husk not so hard and spungye, the seed whereof were devoured by the birds, and the husk remaining on the trees, the leaves whereof were in forme of our houslick in England, but not so thick.

This description could conceivably fit waboom (wagon tree) *Protea nitida*, a characteristic and conspicuous species of the area.

The traveller and historian (and gentleman of the bedchamber to Charles I) Sir Thomas Herbert visited the Cape in July 1627, and described the plants with almost too much enthusiasm, and certainly with more than any previous visitor:

> The Earth abounds with roots, herbs and grasses aromatique, redolent and beneficiall...nature robing the fruitful earth with her choicest Tapistry, *Flora* seeming to dress herself with artlesse Garlands; *Alcinoe* and *Tempe* serving as Emblems to this Elysium.

In 1620 two English commodores got as far as naming one of the foothills of Table Mountain 'King James His Mount' building a cairn, raising the St George's flag and taking possession of the land in the name of the sovereign. The commodores also gave a flag to the Khoekhoen to display to any ships, and left a letter asking the Dutch to treat the inhabitants with kindness as they were now the King's subjects.

The indifference that the English government subsequently displayed to this annexation and to most other proposals to claim the Cape for the Crown might be explained by their preoccupation with their American colonies.

In the meantime, ships kept calling and a clearer and more detailed picture of the Cape began to emerge. Any study of the Cape flora *in situ* by botanists, however, or the selective collection of particular species, notably bulbs, came only with increased visitors and, ultimately, permanent settlement by Europeans in the shadow of Table Mountain in the mid-seventeenth century.

The only exception to this early state of botanical affairs was the visit of Justus Heurnius, who sailed on the *Gouda* to Jakarta where he was to spend 14 years as a missionary. In April 1624, the ship called at the Cape and during the few days Heurnius spent ashore he made sketches of some of the plants he found. After he reached the East Indies, he sent the drawings to his brother, Otto, professor of medicine at Leiden. In due course, they were passed on to Johannes Bodaeus van Stapel (1602–1636), an Amsterdam doctor who was annotating a treatise of the botanical works of the peripatetic Greek philosopher Theophrastus, a pupil of Aristotle.

Unfortunately, van Stapel died before he finished the task, and *Theophrasti eresii de historia plantarum* was completed by his father and published eight years later in 1644. In this, the drawings done by Heurnius at the Cape 20 years previously are copied. The sketches are simple and diagrammatic (certainly when compared with those of, say, Ferrari), but do allow most of their subjects to be identified to the level of genus at least. Our old favourite *Haemanthus coccineus* appears yet again, this time in the guise of *Tulipa promontorii bonae spei* – the 'tulip from the promontory of good hope'. Another recognisable species among Heurnius' drawings is '*Iris uvaria promont. bonæ spei*', now known as *Kniphofia uvaria*, the red-hot poker.

The arum lily *Zantedeschia aethiopica* also made its European debut about this time, featuring in *Recueil des plantes du Jardin du Roi* by Guy de la Brosse (1586–1641), physician to Louis XIII. This work describes the plants growing in the royal garden in Paris and was published post-humously following de la Brosse's death in 1641.

The appearance of the red-hot poker and arum lily, two Cape plants with which today's gardeners are familiar, provided the first real taste of horticultural things to come; the dawn of the golden age of plant collecting at the Cape of Good Hope was not far over the horizon.

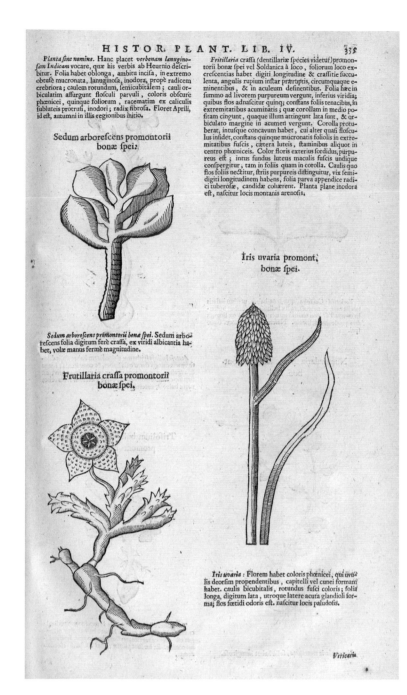

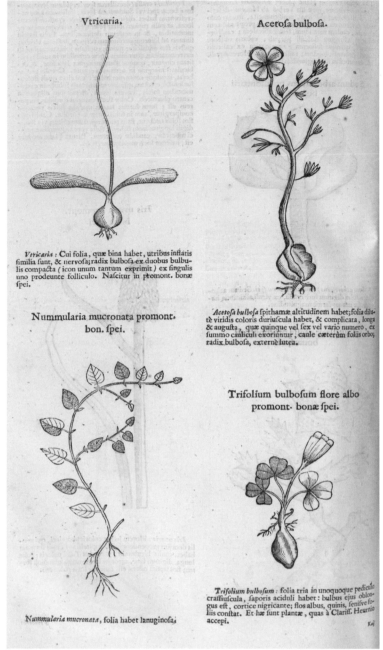

Drawings made at the Cape by Justus Heurnius (1587–?1651) in April 1624 were later reproduced in a modernised version of the botanical works of the Greek philosopher Theophrastus by Johannes Bodaeus van Stapel (1602–1636), completed after his death by his father. A pig's ear *Cotyledon orbiculata* and a red-hot poker *Kniphofia uvaria* are shown together with a carrion flower whose blooms smell of rotting flesh to attract pollinating flies. Labelled 'Fritillaria crassa', the plant was later named *Stapelia variegata* in honour of van Stapel. It is now known as *Orbea variegata*.

On the following page an unidentified bulb is joined by *Oxalis versicolor* (top right), another *Oxalis* (possibly *O. purpurea*, bottom right), and what is possibly *Centella villosa*.

CHAPTER TWO

The Company's Garden

In 1950, Professor R. H. Compton wrote: 'The history of the Europeans in South Africa begins with a garden'. At the time, Compton was director of Cape Town's famous botanical garden at Kirstenbosch, but the one to which he ascribed such historical significance was called the Company's Garden. Today, the site is still a garden, although its function as a green space in the centre of Cape Town has long since moved away from the practical reasons for its establishment within three weeks of the first permanent settlement at the Cape in 1652.

The Portuguese monopoly on the Cape route having been usurped, the English and Dutch established their eastern trading headquarters in India and Java, respectively. In 1611 a new route to the Straits of Sunda was discovered by a VOC captain, using the favourable winds of the south-westerly monsoon. The Cape was midway between Europe and the East Indies on this route and, having also failed to capture Moçambique from the Portuguese in 1609–10, it seemed sensible for the Dutch to establish a station somewhere on the Cape coast. Here they could replenish their supplies of water and fresh meat and also take on green foodstuffs to counter the scurvy that often left the ships seriously short-handed. By this time, indeed, so many ships were stopping off for rest and recuperation at the Cape that they even left messages for one another, notably at the famous 'post office tree', a venerable milkwood *Sideroxylon inerme* not far from Cape Agulhas, Africa's most southerly point.

Despite the increased traffic and demands of the voyage, it was still many years before the VOC approved a proposal:

> to provide that the East India ships, to and from Batavia may procure herbs, flesh, water, and other needful refreshments – and by this means restore the health of their sick – a general rendezvous be formed … at the Cape of Good Hope.

Up to now the VOC considered that the Cape was of little use other than to provide 'water and a little scurvy grass'. A station there would also be expensive to maintain, so its ships were more inclined to refresh at St Helena, particularly on the return voyage. By 1640, however, most of the feral pigs and goats on that island had been eaten.

The ultimate incentive for the VOC to colonise the Cape came from the experiences of the crew of the *Haerlem*, wrecked in Table Bay in March 1647. There were no casualties, but there was not enough room in the accompanying vessel, the *Oliphant*, to take all the shipwreckees back to Holland. About 60 men volunteered, therefore, to stay at the Cape until they could be rescued.

After initial mutual distrust between the Dutchmen and Khoekhoen, more cordial relations were established and cattle and sheep were bartered for. There was also no shortage of wild game. Clean water was plentifully available from the streams flowing down from Table Mountain, the coastal plain was fertile enough to grow crops, and there were some indigenous plants that could be harvested to treat scurvy. All in all, it seemed an ideal place to the extent that when, early the following year, a ship called in with almost 100 sick crew members aboard, enough fresh food and water was available on shore to restore them all to health.

In April 1648, the remaining crew members of the *Haerlem* were rescued and taken back to Holland. One of them, a junior merchant called Leendert Janssen, had kept a journal of his exile and pondered much on the merits of the Cape. In July 1649 he presented the VOC with his 'Remonstrantie', a petition that he entitled 'Exposition of the Advantages

---

opposite

Mixed in with a number of species from Central America, a variety of striking Cape plants are prominent in this work, painted in about 1737 by the Dutch artist Laurens van der Vinne the Younger (1712–1742). Among those that can be identified are, from top left, a yellow mesembryanthemum, arum lily *Zantedeschia aethiopica*, sugarbush *Protea repens*, tree pincushion

Leucospermum conocarpodendron, *Haemanthus coccineus* (bottom left, next to a prickly-pear cactus *Opuntia*), *Mimetes cucullatus*, *Protea scorzonerifolia*, and *Pelargonium inquinans* behind the pink, trumpet-flowered belladonna lily *Amaryllis belladonna*. A puff adder or viper lurks menacingly in the corner – a serpent in the Garden of Eden? *Museum Boerhaave, Leiden*

Until recently, cancer bush *Lessertia frutescens* was known as *Sutherlandia frutescens* after James Sutherland (1639–1719), the first keeper of the Edinburgh Physic Garden, founded in 1677 and later to become the Royal Botanic Garden. A traditional remedy of the Khoekhoen, it was used in the treatment of afflictions ranging from diabetes and dysentery to piles and peptic ulcers, as well as cancer. Modern medical research has yet to identify any clinical effects although it is recognised as a general tonic and a booster of the immune system. The species was well known to us in the Cape of Good Hope Nature Reserve where this specimen was painted.

opposite
Landing of van Riebeeck at the Cape of Good Hope, 1652.
This romanticised painting is by Charles Davidson Bell (1813–1882), the Scots-born surveyor-general of the Cape, and artist and designer of stamps and medals.
National Library of South Africa, Cape Town

to be Derived by the Company from a Fort and Garden at the Cape of Good Hope.' In it he noted how the *Haerlem* crew had not only survived, but thrived at the Cape. It was a temperate area that did not harbour the diseases that so painfully reduced life expectancy in the tropics. Surely, he argued, anything was better than the VOC hospital at Batavia (now Jakarta) which:

> to great cost and injur … is crowded with invalids, who lie there for months without doing any work, but drawing wages notwithstanding? How severely it will be felt by the crews to come home without touching at any place of refreshment, and thus the Company's ships, in the event of great sickness and many deaths, would encounter no small danger.

This was no exaggeration, as the voyage from Europe to Jakarta could take seven months or more. The length and, therefore, cost of the voyage would be reduced if St Helena was bypassed. Janssen also insisted that the native peoples of the Cape were not, as had been previously reported, cannibals, but 'responded to kindness, learnt Dutch easily, and might become servants and be taught Christianity.' Perhaps most importantly, if the Dutch didn't establish a presence at the Cape, then it might be annexed by 'our public enemies the Spanish and Portuguese.'

The VOC were a difficult bunch to win over, but the prospect of losing out to Spain or Portugal, or continuing to fritter away more money on the sick, dying and work-shy was probably just too much for them to bear. So, in March 1651, they reluctantly approved the formation of a 'general rendezvous' at the Cape.

The Cape Colonists

On Christmas Eve, 1651, a fleet of three ships set sail from Texel, bound for the Far East via the Cape of Good Hope. About 90 of the 200 or so people aboard were to remain at the Cape under the command of Jan van Riebeeck (1619–1677). It was no coincidence that van Riebeeck had been on the ship that had rescued the crew of the *Haerlem*. On the voyage home he had tapped them for information and by the time he got back to Holland he was convinced of the potential of the Cape. Bolstered by Janssen's enthusiastic written report, van Riebeeck convinced his superiors of the merits of the Cape and they, in turn, appointed him leader of the colonising expedition.

On the evening of 7 April 1652, van Riebeeck first went ashore from his flagship *Drommedaris,* and two days later he marked out a site for the settlement fort. Cape Town was born.

Van Riebeeck lost no time in establishing a 'refreshment station'. The construction of living quarters and defences was a priority but on 1 May, only three weeks after landing, van Riebeeck recorded that:

> the gardener and his assistant are engaged in sowing some vegetables and other herbs by way of experiment, so as gradually to clear some good land near the moats about the fort; for which purpose more hands will be allowed him as soon as the pressure of work will permit.

The gardener was Hendrick Hendricxen Boom, originally from Amsterdam, and the garden of which he was now superintendent and in which ploughing and planting was soon underway, was known as the 'Company's Garden'.

Despite his enthusiasm and hard work, not everything went the gardener's way. On 14 June, van Riebeeck recorded that all their seedlings had been destroyed by rain and hail storms. These conditions did, however, encourage the growth of wild plants, and less than a week after the destructive deluge, the commander noted sanguinely that:

> Many [indigenous] herbs spring forth as a result of the rain and we expect much success from our Dutch fruits during the prosperous season.

The rest of the winter proved wet and cold and, although many of his plantings were destroyed, some survived and he was later able to report that 'We have . . . successfully cultivated Dutch greens that for our table and the such we can avail ourselves daily of radish, lettuce and cress'. Van Riebeeck also noted that, while much of the seed was washed away, it was later found 'sprouting nicely in various places; it is a pity that they were not allowed to stay in their proper places.' Learning from their experiences of 'the largest and heaviest showers in the world', the colonists constructed dykes and ditches around and through the garden and further flooding was averted.

Having survived the north-westerly gales and downpours that typify winter at the Cape, the colonists and their crops then had to put up with the other unavoidable climatic feature that afflicts this corner of the continent – the south-easter. This wind blows with formidable strength and tedious persistence for much of the summer from November to February:

> We had a busy time securing our dwellings with stays; and nearly all our Roman beans and many peas, laden with pods and blossoms, have been blown down and the wheat was struck flat against the ground during those strong winds.

The south-easter struck again later and the:

> wind blew so fiercely that all our beans, peas and barley which already showed very nice fruits and ears, were torn to pieces, so that it is feared that hardly anything that grows high will ripen properly, the southeaster blowing so hard that nothing can keep upright.

By the end of 1652 the colonists had already experienced pretty well everything that the Cape climate could throw at them. Given these circumstances, it is hardly surprising that the indigenous vegetation of the Cape is well adapted to withstand chilly winter flood, searing summer drought, and gales from one direction or another all the year round. Despite the weather, van Riebeeck was resourceful and energetic, and the settlement soon began to prosper.

During his sojourn, van Riebeeck appears either to have studiously ignored the indigenous wildlife or to have had insufficient time to study it. In his journal, or *dagboek*, only occasional reference is made to the Cape's natural history. Large animals are mentioned when they damage crops, terrorise the inhabitants, or end up on the dinner table. He makes virtually no mention of any flowers of unusual qualities or appearance, confining his botanical observations almost exclusively to plants of practical use.

Most useful to the settlers were the forests 'so abounding with big, tall and straight-growing trees, that one would be able to get entire masts for ships there by the thousands'. Van Riebeeck was delighted to

The winter of 1652, from July to August, was an exceptionally severe one; part of Table Mountain was white with snow, and so much rain fell that the land resembled a sea. There was much sickness and several deaths among the garrison. But all the privations and discomforts of these months were forgotten with the advent of the Cape spring, in September, when the verdure of the hills and valleys afforded pleasure to the eye.

John Noble, Discovery and Early History of the Cape, in *Cape of Good Hope Official Handbook* 1886.

find this bountiful supply of timber on some of the slopes and in the deep gullies of Table Mountain, particularly at Hout (Wood) Bay. Predictably, felling took place on such a scale that within only three years of arriving van Riebeeck was compelled to decree that the use of one particularly valuable species, yellow-wood *Podocarpus latifolius*, should be restricted to sawing for planks as the trees were 'recklessly cut down and destroyed by everybody without consideration.'

Toxic Coffee

One marginally edible crop that van Riebeeck came across was the fruit of the wild almond *Brabejum stellatifolium*, an unusual member of the protea family. On 7 April 1654, on the hillside about a mile and a half from the settlement fort, van Riebeeck reports how his party:

> found trees in profusion, even whole groves full of ripe bitter almonds, apparently fit for feeding up pigs, similar to acorns, which these almonds almost completely resemble in taste. They were much gathered and eaten by the natives, who first peel them, then dry them in the sun for a few days and finally roast them on the fire; we will try this at our earliest convenience.

This he did, and three of the colonists died as a result. This was because wild almond fruits contain prussic acid which, if not broken down or extracted before consumption, can be fatal. A later attempt at making something useful from the abundant nuts was to adopt the technique of the Khoekhoen and dry and roast them, submerge them in running water for several weeks to flush out the acid, dry them again, grind them up and, finally, pour boiling water on them. Thus did the Dutch discover a substitute for coffee.

A more practical and straightforward use for wild almond was recorded in a resolution of 7 November 1659, by which all the land occupied by the Company was to be fenced-in against wild animals and marauding natives. (This was a somewhat more realistic plan than an earlier proposal to make the Peninsula into an island by digging a canal from Table Bay to False Bay). Wild almond fruits, wrote van Riebeeck:

> can be collected in abundance about the end of March or the beginning of April, when they are ripe, so that they can be put into the ground with the first wet season which is at hand … In four or five years time they would form to all appearance a good thick and strong fence, for it has been found that the said bitter almond trees grow as luxuriantly as any willow in the fatherland, and giving thick wood, would be difficult even for human beings to penetrate when intertwined with … thorn bush, and still more difficult for cattle to break through.

The hedge was planted in July 1660. Its wild almond component forms a massive rootstock from which new shoots sprout if the old ones are coppiced or burnt. Such is their resilience that fragments of van Riebeeck's hedge still survive.

Jan van Riebeeck left the Cape in 1662. He never returned to Holland, but sailed with his family to Batavia where he was promoted and continued to work for the VOC until his death in 1677. By the time he left the Cape, about 30 Dutch ships were stopping each year and the settlement had extended well beyond the Company's plan for of a rigidly controlled, tightly economic and strictly utilitarian 'Tavern of the Seas' for its fleet. The importation of African and Malay slaves, the arrival of more European colonists and the granting of free status to many of the original 'burghers'(citizens), who were nominally under the social and economic command of the VOC, radically altered the social structure and physical boundaries of the settlement.

Despite this increase in population and visits, relatively little attention appears to have been given to the wild flowers. It is hard to believe that

Bulbs are hardy, uncomplaining items that can survive long periods of dormancy, springing into life when provided with water. Easy to collect, store and transport, it is not surprising that they formed a large proportion of the plants brought to Europe in the early years of botanical exploration of the Cape. All these bulbs, in the broadest sense of the word, continue to be widely grown in gardens and greenhouses around the world today.

Left, *Eucomis autumnalis*. Right, top, chinchernichee *Ornithogalum thyrsoides*; right, *Agapanthus*. Far right, belladonna lily *Amaryllis belladonna*.

the floriferous proteas or the fabulous spring displays of bulbs and annuals on the sandy flats beneath the hills had failed to make an impression on the colonists but, if they did, there was no mention of it. Although van Riebeeck's *dagboek* provides one of the earliest accounts of some aspects of the natural history of the Cape, it was for later visitors to begin to appreciate its special qualities.

Good Soil in a Fruitful Garden

Having been afflicted with scurvy on the passage from Holland to the East Indies, Johan Schreyer (dates unknown) was left to convalesce at the Cape in 1668. Once recovered, he found the place so much to his liking that he stayed for six years. The 'Cape Resolutions' record that in 1670 he was promoted to Under-Surgeon 'having been continually employed as a surgeon with our inland barter-parties'. He was later promoted to 'Upper-Surgeon' and put in charge of the hospital.

Schreyer made a significant contribution to the documentation of life at the Cape at this time, particularly in his detailed first-hand accounts of the customs of the Khoekhoen. In his book *Neue Ost-Indiansche Reisz-Beschreibung 1669-1677* he makes a brief but tantalising reference to plants:

> I also brought some boxes of bulbs with me to Holland, but left them with an eminent man in Seeland [a province of the Netherlands]. There also I learned that, although these were planted in a very good soil in a fruitful garden, they did not by a long way produce such lovely flowers as they were wont to do in their native soil.'

Plants, like goldfish, are perverse organisms – if you're too nice to them they die. Most western Cape species have evolved to grow in conditions of chronic nutrient deficiency and seasonal drought that other plants would find distinctly alarming. 'Very good soil in a fruitful garden', therefore, would be an almost sure-fire way of killing off your precious Cape collection. Nonetheless, this casual observation by Schreyer hints that far more was happening with Cape plants in Europe than the rather sparse records emanating from their home ground would suggest. This is supported by the appearance in 1675 of a number of species in a short publication containing descriptions and drawings of plants collected at the Cape by Paul Hermann.

Born in Saxony in 1646, Hermann's botanical interest almost finished as soon as it had started when, at the age of ten, he nearly drowned while gathering plants on a river bank. Having survived that early near-disaster, and after studying medicine at the Universities of Leiden and Padua, he was appointed medical officer in Ceylon (now Sri Lanka) by the VOC. One of the prime movers behind his commission was Willem Bentinck, an enthusiastic aristocratic horticulturist who specialised in growing rare plants on his estate at Zorgvliet near The Hague.

On the way to his new posting in 1672, Hermann made a stopover at the Cape and collected some plants, a few of which he gave to a ship's

opposite
One of the most common bulbous plants around Jan van Riebeeck's settlement would have been *Moraea flaccida*. Being poisonous, it is avoided by livestock and proliferates where other plants have been over-grazed. Popularly known as Cape tulip or by its older botanical name of *Homeria*, it has been grown in Europe for over 300 years and, unlike the majority of plants in cultivation, remains all but unchanged from its ancestral species.

above
The first systematic collection of Cape plants was made by Paul Hermann (1646–1695). His herbarium was put up for sale in 1711 and the 'Plants gathered by Dr Hermann at the Cape of Good Hope in 1672' were ultimately acquired for Sir Hans Sloane and are now in the Natural History Museum, London. Among the remarkably well preserved specimens are wild dagga *Leonotis leonurus*.

© *Natural History Museum, London*

surgeon whom he met in Cape Town. Returning to Europe, the surgeon, in turn, handed the plants over to Thomas Bartholinus, a physician and anatomist who lived in Copenhagen. (At this time, Dutch vessels were forced to divert to Denmark because of the English blockade of Dutch ports during the Third Anglo-Dutch War of 1672–74). Three years later the plants appeared under the heading *Plantae novae Africanae* in Bartholinus's journal *Acta Medica et Philosophica Hafniensa*. Although only nine species were illustrated and a further ten enumerated, these few pages represent the first account devoted exclusively to flowers from the Cape. The plants include a heath, an iris, a shrub, and a straggly annual, all of which can be found on and around Table Mountain, so Hermann probably didn't go far to gather his trophies. Nevertheless, he can be credited with making the first 'official' herbarium collection of Cape plants.

Having been appointed director of the University Garden at Leiden, Hermann returned from Sri Lanka to Holland in 1680. He became one of the leading botanists of his day and amassed a collection of plants from the East Indies, the Americas, and the Cape. His 1687 catalogue of the species growing in the University garden includes 34 Cape species while his *Paradisus Batavus*, published posthumously in 1698, contained some additional ones.

At the time of his death in 1695, Hermann was working on his *Prodromus plantarum Africanarum*, but it was not until 1737 that this work appeared in print. A 'prodromus' is normally the forerunner or introduction to a larger work, but in this case it took the form of an appendix to *Thesaurus Zeylanicus* by Johannes Burman, professor of Botany at the University of Amsterdam. It listed 791 Cape plants that had been accumulated by Hermann over the years, the majority no doubt courtesy of the VOC.

Bulbs the Size of a Human Head

In the spring of 1673, Wilhelm ten Rhyne, also a physician with the VOC, made a three-week stopover at the Cape. He describes the Company's Garden, as:

> a lovely sight with its plantations of lemons, citroens and oranges, its close hedges of rosemary and its laurels... It is the very essence of greenness set in the midst of thorns and barren thickets.

It is more than likely that 'thorns and barren thickets' were the product of overgrazing and stripping of the veld by the colonists. The area can

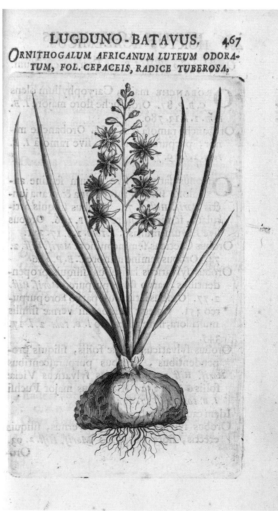

look a bit parched, but not usually at the season when ten Rhyne was there. As if to contradict himself, he then goes on to describe how:

> this soil, barren though it may be, abounds in plants of every sort...comely ericas, house leeks of various kinds, ornithogalums and narcissi. The bulbs of these often equal a human head in size, and specimens were long ago sent to Holland to satisfy the curious. There are also geraniums which smell sweetly at night.

The latter would be our old friend *Pelargonium triste*.

Ten Rhyne made sketches of some Cape plants and sent them to Jacob Breyne, an ardent plant and botanical art collector. Illustrations based on some of these subsequently appeared in the latter's *Exoticarum aliarumque Minus Cognitarum Plantarum Centuria Prima, cum Figuris Aeneis Summo studio elaboratis*, being 'The first hundred exotic and other lesser known plants, with added figures and observations', published in 1678. This treatise features 48 depictions from a list of 85 Cape species received from Hermann, other collectors in Holland and directly from the Cape itself.

A further legacy of Jacob Breyne, who died in 1697, is a valuable portfolio of Cape flower paintings, some of which are likely to have been executed in Holland from specimens growing in gardens there, while others were probably painted *in situ* at the Cape, perhaps in the Company's Garden. The frontispiece portrays a spectacular arrangement of blooms including gladioli and daisies. Even in an age when new and unusual plants were arriving by the bucket-load from different parts of the world, those from the Cape must have made a particularly strong impression on the botanists and horticulturists of the day.

During its formative years the Company's Garden, together with other plots and small fields established in the vicinity of van Riebeeck's fort, was used exclusively for cultivating fruit and vegetables to feed the increasing resident population and the crews of visiting ships. After a while some settlers, including Boom the gardener, were granted the status of 'free burghers'; as such, they were allowed to grow their own produce to sell to the Company. This allowed the Company's Garden to be used on a more experimental basis, testing and trialling various crops to assess their potential for use at the Cape. Numerous ornamentals were introduced and, most importantly in our context, the garden was also used as a holding pen or propagation unit for native plants destined for Europe.

Today, the original Company's Garden, also known as the Municipal Botanic Garden, is flanked by an avenue of oaks and set amongst the handsome buildings of the South African Museum, Library, and Parliament in central Cape Town. It occupies much of the site of the original and contains a garish and luxuriant variety of predominantly exotic tropical plants from the Americas and East Indies. Only by turning your gaze towards Table Mountain are you reminded that the real botanical riches of the Cape are to be found over the garden fence.

The Soil of this Country is of a brown colour; not deep, yet indifferently productive of Grass, herbs and trees. The Grass is short, like that which grows on our Wiltshire or Dorsetshire Downs. The trees hereabouts are but small and few; the Country also farther from the sea does not much abound in trees, as I have been informed. The Mould or Soil also is much like this near the Harbour, which though it cannot be said to be very fat, or rich Land, yet it is very fit for cultivation, and yields good crops to the industrious Husbandsman …

William Dampier, 1691.

above
One of the Cape plants featured in Hermann's catalogue is '*geranium africanum*'. This is *Pelargonium gibbosum*, a relatively early arrival in Europe and cultivated in England by at least 1712. It is sometimes known as the Gouty Pelargonium because of the swollen stem-joints that provide support and may hold reserves of moisture. This one clambered purposefully through the shrubs of our Cape Peninsula garden as if on some sort of quest. Its small flowers are night scented ('*noctu olens*') and pollinated by moths.

CHAPTER THREE

God and Friendly Nature Combined

In the 17 years following Jan van Riebeeck's departure there were no fewer than ten commanders (known as governors after 1691) of the Dutch settlement at the Cape. The appointment of one of these, Simon van der Stel (1639–1712) in 1679, saw a sea change in the extent of knowledge and recognition of the Cape's flora and natural history in general. By this time, the infrastructure and functioning of the community were well consolidated, allowing time and resources to be devoted to exploring the lie of the land and more of an interest to be taken in what was found there.

Van der Stel's first and most enduring legacy at the Cape was to establish a new settlement only a few weeks after his arrival. Returning from a tour of the outlying policies, he took a fancy to a spot beneath the mountains on the eastern side of the Cape Flats. With proprietorial self-concern and posterity in mind, he named this beautiful and fertile place Stellenbosch.

The Distinguished Botanophile

An enthusiastic agriculturalist and explorer, van der Stel was also keen on his plants and was directly or indirectly responsible for significantly increasing the volume and variety of material sent back to Holland. A century later, he was hailed by Johannes Burman as 'the distinguished botanophile', so he clearly made his mark. Although described as 'loyal, honest and industrious', van der Stel lacked the diplomatic skills necessary to endear himself to his masters in the VOC being, it was

claimed, 'haughty and inaccessible'. It was also reported that he was inclined to spend more time improving his own estate and vineyard at Constantia than he was on Company business. This resulted in him being dismissed in November 1699, to be replaced by his son, Willem. By 1707, however, it was obvious that as far as the van der Stels were concerned it was too much a case of 'like father like son' as Willem was sacked for 'misconduct' in the form of misappropriation of Company time, labour and materials to develop his personal estate, together with an arrogant attitude towards the free burghers that did not enhance his position as governor. So back to Holland he went. His father remained at the Cape and died there in 1712.

As part of the Company's objective to make as much money as possible from any of the Cape's natural resources, various expeditions had been sent north from Cape Town to look for the gold and other valuable commodities that were rumoured to be there. A number of expeditions, including six between 1660 and 1663, failed to find anything of note. In 1682, Ensign Oloff Bergh led a party that made the deepest incursion yet to Namaqualand, but also found nothing of value.

Frustrated by this failure, but tantalised by the copper artefacts that itinerant tribesmen brought with them from the north, van der Stel took command of his own expedition to find the source of the copper, rumoured to be in the mountains of Namaqualand.

The expedition left Cape Town in August 1685 and included almost 70 personnel, a carriage and six horses for the governor, eight carts, seven wagons (one transporting a boat), and 289 draught oxen. Although primarily a commercial venture, van der Stel was instructed to record 'the country, mountains, rivers, roads, and the people, as also of timber and forests … like wise of everything that is perceived as worthwhile.' The inclusion of Hendrik Claudius (c.1655–c.1697), a medical botanist and artist, on the expedition could thus be justified.

The trip lasted four months, and although copper ore was found it

opposite

Carolus Linnaeus (1707–1778), founder of the refreshingly simple and enduring binomial system of naming plants and animals. He is pictured here in traditional Lapp costume and clutching his favourite flower, the only species named after him,

twinflower *Linnaea borealis* of northern Europe and America. Linnaeus was captivated by the plants that he received from the Cape and it was one of his few stated regrets that he never travelled there himself. *By permission of the Linnean Society, London*

can be described as crude), are accurate and detailed enough to allow the identification of a good proportion of their subjects.

Claudius's drawings were later extensively copied by himself and other botanical artists. The earliest record of this comes from Father Guy Tachard who called at the Cape in June 1687 as a member of a French mission to Siam that included 14 Jesuit mathematicians. One member of the company, Father de Bèze, records how he went:

> in search of rare plants, or to make notes on others. I have found a great many, and some beautiful ones. Although it is winter here, the land is decked with flowers as are our fairest meadows in the month of May.

This was at least an improvement on a previous observation that:

> We were greatly surprised to find one of the most beautiful and curious gardens [the Company's] which I have ever seen, in a country which appears to be the most barren and miserable in the world.

Tachard and de Bèze met with van der Stel, who:

> spoke to us of some curious plants which he had found on his travels, and shewed us a collection of them. He was good enough to allow us to have drawings made of the more uncommon ones.

These were most likely executed by Claudius and appear in Tachard's publication of 1689 describing his second of three journeys to Siam. Of Claudius, Tachard records how it was he:

> whom the Dutch maintain at the Cape on account of his ability … and as he draws and paints to perfection both animals and plants, the Dutch keep him there in order to assist them in the exploration of new regions and to work at a natural history of Africa. He has already completed two thick folio volumes of various plants painted from nature, and has collected specimens of all kinds … As this learned doctor had already made several journeys, to a distance of one hundred and twenty leagues north and east of the Cape to make new discoveries, it is from him that we obtained all our knowledge of the country. He gave us a little map made by his own hand, and some drawings of the inhabitants and animals which I am inserting in my book.

Claudius was, indeed, enthusiastic and generous with imparting information. Too generous, in fact, as he soon found himself packed off to the Far East by Governor van der Stel who did not like subordinates receiving any credit (as Claudius did in Tachard's account when it was published). There was, in any case, deep residual suspicion of the French and their motives for calling at the Cape and any fraternising beyond civility was not encouraged.

Arguably the most famous subsequent appearance of Claudius's watercolours, or at least copies of them, is in the three-volumed *Codex Witsenii*. Drawn from the originals at the Cape in 1692, they were commissioned by Nicolaas Witsen, a noted traveller and geographer who was soon to become a director of the VOC.

A varied career in diplomacy, trade and local politics allowed Witsen to travel and to befriend and correspond with fellow natural-history

was not in sufficient quantities or of sufficiently high yield to make mining and transporting it economically viable. The expedition members also failed to register the fact that, at one stage, they were crossing the wide plains that hold some of the world's richest alluvial deposits of diamonds.

The expedition was too late in the season to witness the spectacular floral displays of Namaqualand. Of value to botanical historians, however, are the illustrations made by Claudius of some of the plants and animals that were found. At least one copy, if not the original, of van der Stel's journal of the expedition was sent to the VOC in Holland. This included the Claudius watercolours which, if of limited artistic merit (indeed, they

enthusiasts. He was a friend of the van der Stels and it was through one of them, most likely Simon, that he obtained the illustrations that form the Cape component of the *Codex Witsenii*.

A 'codex' is an unpublished manuscript or collection of paintings. Witsen's reportedly contained some 1,500 paintings of plants from the Cape and the Far East. After his death in 1717, it made its way to Caspar Commelijn, thereafter to Johannes Burman. The latter included 97 Witsen illustrations in his *Rariorum africanarum plantarum*, published in Amsterdam in 1738–39. Other Cape plants in this volume were drawn from specimens growing in the Amsterdam Athenaeum garden which was probably the source of some of the most botanically accurate and sophisticated illustrations, at least when compared to those of Claudius.

In 1691 the Bishop of London, Henry Compton, accompanied William II to Holland. There he met Paul Hermann and received a copy of the *Codex Witsenii*, this particular volume later becoming known as the *Codex Comptoniana*. A number of its illustrations were copied and adapted by the English botanists Leonard Plukenet and James Petiver to include in their botanical works *Phytographia* (1691–1705) and *Gazophylacii naturae & artis* (1702), respectively. A significant mileage was, at the end of the day, obtained from Claudius's rather simple but significant watercolours.

Mr Donker and the Widow Oldenland

In December 1690, the governors of the VOC wrote to van der Stel, advising him that:

> There should be with you in the Garrison a certain Hendrick Bernard Oldenland of Lubeck, a very good botanist or connoisseur of herbs who studied medicine for three years at the University of Leyden. Your honour would do well in appointing and employing him to grow and collect any medicinal herbs and plants which might be found or discovered with you, and which could be made use of, so that Batavia and Ceylon could be supplied with them to meet their requirements.

Oldenland had recently bought his freedom from the Company and set himself up as an agricultural consultant, but van der Stel persuaded him to rejoin the payroll and he was appointed master gardener and land surveyor in 1692 or early '93. He was not to enjoy his position for long, as he died in 1697. He did, however, make a collection of indigenous medicinal plants and sent specimens and material back to Holland.

Some years later, Oldenland's *Herbarius Vivus* was examined in Holland by François Valentijn, a Dutch minister and historian to the VOC. The collection, according to Valentijn, consisted of:

> 13 or 14 volumes in folio, with a nice description in Latin of each plant. I found this work, left by him, in the year 1714, and of which I have occasionally perused with great delight...The plants were unusually fine, exceedingly well dried, and still of such a lively colour that it was a treat to see them.

Valentijn also relates how the collection was 'greatly desired' by some Englishmen but the owner, a Mr Donker, the third husband of the widow Oldenland, 'required from them far too much money for their liking'. Valentijn was clearly taken with the Cape flora and observed that:

> There is nothing one should marvel at more than the whole fields with flowers which grow here wild and which are of a strong and beautiful colour that it is a pity that all these could not be drawn from life, and in their natural colours, by a competent artist.

The impact of Oldenland's botanical efforts ranged well beyond the Cape and even Holland. Plants that he sent back, together with his personal collections, ultimately came to the attention of many of the leading botanists of the day. A collection reached Johannes Burman, and Oldenland or his widow sent specimens to Petiver that later featured in his *Hortus Siccus Capensis* (literally 'dried plants of the Cape', but

JOANNES BURMANNUS Med: Do
Botan: Profeſſ: Amſtelæd: anno aetatis trigeſimo.

Non haec effigies BURMANNI sculpta; sed ipsum
Arte Promethea vivere in aere vides.
Talis Paeoniis meditatur pellere morbos
Viribus, aut laetas Chloridos auget opes.
His oculis proavi, genitorque hac fronte renident.
Nil non egregiae pignora gentis habet.
Gaudeat Amstelidum longos Hygiea perannos
Huic tabulae carum composuisse caput.
JAC. PHIL. D'ORVILLE.

above

Johannes Burman 1707–79 succeeded Caspar Commelijn as professor of botany at the Amsterdam Athenaeum. This engraving is from his *Rariorum africanarum plantarum*, published in 1738–39. Over 200 Cape plants are depicted, some drawn from life in Dutch gardens. Others were copied from existing works, notably the Witsen Codex received from Commelijn's widow and which inspired in Burman 'a return to the former delights of Flora' following a protracted illness.

opposite

Many plants in Jan Commelijn's garden were described in *Horti Medici Amstelamodensis*, a book that he was working on at the time of his death and which was finally completed by his nephew, Caspar (?1667–1731). The engravings were adapted from watercolour studies by Johan (?–1673) and Maria Moninckx (unknown dates), and include a number of Cape species, such as red-hot poker *Kniphofia uvaria* and the blue *Agapanthus africanus,* that have since become familiar garden plants. The illustrations in the Kew edition of this work are superbly handcoloured; other editions remain unpainted.

essentially meaning 'Cape Herbarium'). Petiver worked as a demonstrator at the Chelsea Physic Garden and as an apothecary from premises in Aldersgate Street, London. These were conveniently positioned for Petiver to be first in the queue for returning ships to see if they had any interesting bits and pieces to add to his collection.

The northerly flow of plants in the closing years of the seventeenth century generated a preoccupation with the Cape flora in Holland and other European countries. Simon van der Stel would like to have had the credit for this, but is recognised as being a bit of a schemer and quite happy to accept kudos where it was not entirely due. There seems no doubt that he was responsible for reinvigorating the Company's garden, but the actual collecting and cultivation were done by the likes of Oldenland and his successor Jan Hartog. Through their efforts, van der Stel could keep up the supply to Holland, as was noted by Georg Meister. On calling in at the Cape *en route* from the Far East in early 1688 he was charged with the task of transporting:

> three chests for HRH the Prince of Orange; five chests for HE Caspar Fagel, Pensionary of Holland; and nine chests of such trees, flowers and garden plants for the Hortus Medicus at Amsterdam, in all seventeen chests with soil and all sorts of plants.

Meister made detailed observations of the people at the Cape. His accounts of wildlife, while appreciative, are sparse, as he admits:

> I say but little of the lovely green meadows and fields adorned with a thousand strange colours of flowers, where God and friendly Nature combine to show a masterpiece...if there is anywhere in the world where Nature plays incomparably with rare and lovely colours of her flowers and herbs, it is on this extreme point of Africa, the Cap bon Esperance.

Some of van der Stel's plants were also sent to Paul Hermann at Leiden, but the majority went to Johannes (Jan) Commelijn and his nephew Caspar at Amsterdam where they were in charge of the Hortus Medicus, the medicinal garden. This had been established by the elder Commelijn and Joan Huydecoper, the mayor of Amsterdam, in 1682.

In addition to collecting and cultivating, Jan Commelijn devoted himself to an account of the garden's plants. He died in 1692 and sadly did not see this in print, but the two volumes were completed by Caspar and published in 1697 and 1701 under the title *Horti medici Amstelodamensis*.

The works were illustrated with hand-coloured engravings taken from original artwork by Johan and Maria Moninckx. A variety of South African species, including crassulas, aloes, daisies and lobelias, are beautifully depicted. Of particular interest are plants such as *Sansevieria hyacinthoides*, a species that occurs in the eastern Cape

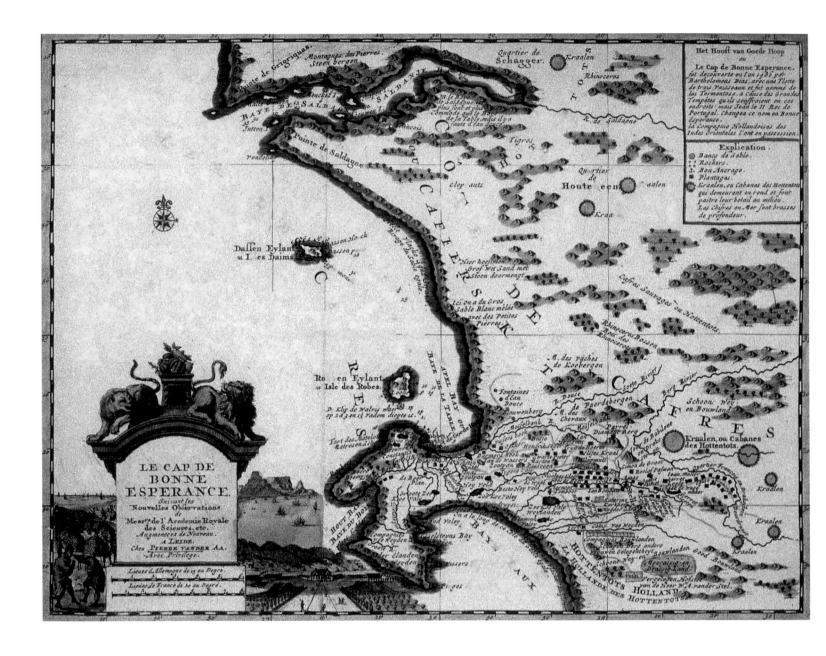

some 600 kilometres from the Cape of Good Hope. This had been the destination of an expedition in 1688 that saw Oldenland, under the command of Ensign Isaq Schrijver, venture further across the country than any European had done before. Many of the plants brought back for the Company's Garden were succulents such as aloes and, perhaps, this *Sansevieria*. It's not known if Oldenland christened it 'Mother-in-Law's Tongue', or if this accolade was reserved for its subtropical relative, *S. trifasciata* which, somewhat inexplicably (at least on aesthetic grounds) was to become such a popular houseplant.

Tulbagh and the Father of Taxonomy

If the last years of the seventeenth century were noted for the new and exciting plants coming from the Cape, the first few of the eighteenth were, by contrast, rather quiet. This was, however, a period of significant development and expansion of the Cape colony.

Plant collecting began to recover its momentum with the rising through the ranks of Rijk Tulbagh (1699–1771), who had arrived at the Cape as a junior cadet in 1716 and by 1751 had been appointed governor. In this position he was able to indulge his abiding interest in natural history, despite having had an eye blown out by a gun that exploded in his face when he was attempting to secure some specimen of bird or mammal. During his 20 years in office Tulbagh sent many consignments of seeds, bulbs, living plants and dried specimens back to Holland, primarily the botanic gardens at Leiden.

Tulbagh regularly dispatched his staff on collecting trips and organised large-scale expeditions to Namaqualand in the north and Caffraria to the east. The indigenous species component of the Company's Garden was thus increased, enabling it to become less a vegetable plot and more a botanical garden. Its superintendent, Johann Andreas Auge (1711–1805), was responsible for sorting and packaging the material collected on the expeditions. He was also not alone in supplementing

The genus *Tulbaghia* was named by
Linnaeus after Rijk Tulbagh (1699–1771),
governor of the Cape for 20 years and
an enthusiastic botanist who despatched
many consignments of plants back to
Europe. *Tulbaghia alliacea*, which smells
sweetly at night, here flanks the more colourful *Moraea* (formerly *Homeria*) *elegans*.

Good Hope: that land which no botanist ever before had trod. Oh Lord! How many, how rare and how wonderful were the plants that on this very day presented themselves to Hermann's eyes! In a few days Hermann alone and by himself discovered here more African plants than all botanists who ever before him had made their appearance in the world.

Linnaeus continues:

> Mountains and slopes were covered with succulents … Protea and Leucadendron species made the woodlands shine with silver and gold … From here Hermann sent to his Floral headquarters more new plants than anyone before him, and to this very day the gardens of Europe are embellished by them. Thus did this eminent discoverer acquire eternal fame …

Linnaeus classified pretty well anything that he could lay his hands on according to his new system, providing a practical and concise account of the 7,700 species of plants and 4,400 species of animals known at the time. The first edition of his classification, *Systema Naturae*, appeared in 1735; the last edition, published in 1766–68, ran to 2,500 pages. The tenth edition, of 1758, became the standard reference on the binomial system of classification and remains the benchmark for taxonomic nomenclature today.

Among the thousands of specimens of plants, birds, mammals, fish and insects that Linnaeus received from his various contacts, were a multitude from the Cape; organisms as diverse as the belladonna lily *Amaryllis belladonna*, leopard *Panther pardus,* and stonechat *Saxicola torquata*, have the letter 'L' (denoting Linnaeus as the authority who applied the name) after their scientific names where they appear in a formal context. Linnaeus named 780 species of plants from the Cape Peninsula alone, many of which were sent to him by Tulbagh from 1761 onwards.

The recipient was so impressed with what he had obtained that he wrote to Tulbagh the famous lines, now popularly quoted (with some justification) by those of us interested in promoting the natural wonders of the Cape:

> May you be fully aware of your fortunate lot in being permitted by the Supreme Disposer of Events to inhabit, but also to enjoy the sovereign control of that Paradise on Earth, the Cape of Good Hope … Certainly

if I were at liberty to change my fortune for that of Alexander the Great, or of Solomon, Croesus or Tulbagh, I should without hesitation prefer the latter.

In recognition of his good works, Linnaeus named a genus of amaryllid (lily-like bulbs) after Tulbagh, of which one species, the purple-flowered *Tulbaghia violacea*, has become quite a popular garden ornamental. A handsome butterfly also bears Tulbagh's name as does, perhaps not least, a beautiful village nestling in an idyllic mountain valley north-east of Cape Town.

Linnaeus himself never travelled beyond a few European countries, despite having the opportunity to do so, and it is a great pity that he never visited the Cape as he would have revelled in its riches. Perhaps another of his musings might account for his reluctance:

> Good God. When I consider the melancholy fate of so many of botany's votaries, I am tempted to ask whether men are in their right mind who so desperately risk life and everything else through the love of collecting plants.

Despite Linnaeus's misgivings, in the second half of the eighteenth century, men of questionable sanity became increasingly frequent visitors to the Cape.

opposite
Distinctly less fragrant than its cousin, *Tulbaghia violacea* is known as wild garlic for good reason. Nevertheless, it is becoming an increasingly popular garden plant, especially the variegated cultivar 'Silver Lace'. As with other attractive but odoriferous items (crown imperial *Fritillaria imperialis* being a prime example) it should be planted in full view but as far away as possible in your garden or, preferably, in a neighbour's.

above
The orchid *Disperis capensis* is my favourite Cape flower. Its elegance and charm belie a wiry toughness that allows it to withstand gales and downpours when it comes into bloom in mid-winter. The Afrikaans name 'moederkappie', means 'mother's bonnet'.

CHAPTER FOUR

Men of Questionable Sanity

These were exciting times in the history of botanical discovery at the Cape. Here, the arrival of impassioned collectors and more scientifically-minded botanists was driven by an upsurge of interest in exotic flora in Europe. This was especially the case in England where the gardens at Kew had embarked upon a path that took them from a royal pleasure ground to a place of learning as well as beauty.

This journey began in 1721 when Richmond Lodge in Surrey became a favourite residence of George II and Queen Caroline. A few years later their son Frederick Prince of Wales, and his wife, Augusta, leased Kew House, also known as the White House, next door. This property 'had hitherto been held in high repute as a place where advanced and refined horticulture had long been practised.' The couple made many alterations to the landscape and gardens until in 1751 Frederick, the heir apparent to the throne, died, having seemingly caught a fatal chill while watching some trees being planted.

A member of his household, Lord Bute, soon became the Princess Dowager's personal adviser. Some five years earlier he had been appointed tutor to her eldest son, George. With such close royal connections and George's early ascent to the throne following his grandfather's death in 1760, Bute's career flourished, culminating in his appointment as Prime Minister in 1762. A great lover of plants and an accomplished botanist (he later wrote *The Tabular distribution of British plants* (1787), Bute was also able to advise on the development of the garden at Kew.

In 1759, meanwhile, Augusta had appointed William Aiton as head gardener. Originally from Carnwath in Lanarkshire, Aiton had moved to London in 1754 to train under Philip Miller of the Chelsea Physic Garden.

Around the same time, the architect William Chambers was commissioned to construct various buildings and ornaments around the policies. With all this activity and the development of the estate promoted by the liberal-minded Augusta and steered by Bute, Aiton and Chambers, 1759 is the year generally regarded as the one in which Kew Gardens officially came into being.

All was not rosy in the garden, however. Bute's real or perceived increasing closeness to the widowed Augusta became the target of scurrilous reports in the press and the object of satire. Rumours of an affair between them were almost certainly unfounded, but that didn't stop the gossip and Bute became, it seems, 'the most hated man in Britain.' George III liked neither Bute's political ambitions nor his friendship with his mother, so Bute had no choice but to resign as Prime Minister after only a year, retreating to one of his country estates.

Lord Bute's input to the development of Kew thus ground unceremoniously to a halt, which was a shame as he doubtless made an important contribution to the gardens in their early days. It may be some consolation to him that his name lives on in the plant world in the names *Butea* and *Stuartia* (*Stewartia* of Linnaeus). He was also credited with having 'the most elegant legs in London.'

In 1772, the year of his mother's death, George III combined the two halves of Kew into one estate. His wife, Charlotte Sophia of Mecklenburg-Strelitz, embraced the development of the garden with

opposite

King George III (1738–1820) in his coronation robes, by Allan Ramsay (1713–1784). George first met Charlotte Sophia of Mecklenburg-Strelitz (1744–1818), on 8 September 1761 and married her later the same day. The garden at Kew doubtless provided the queen with a welcome diversion from domesticity and children (she had 15). Charlotte died at Kew Palace, known at the time as Dutch House. George died just over a year later, his madness and associated blindness and virtual deafness having rendered him incapacitated for a decade. In his active years, there is no doubt that, despite political weaknesses, 'Farmer George' made a significant and lasting contribution to the advancement of the natural sciences, including botany.

©*National Portrait Gallery, London*

an enthusiasm that seemed to infect not only those in her immediate circle, but also the nation at large. Within a few years Kew became the prime destination for new plants from around the world, their collectors and couriers seeking to impress Her Majesty and add to their personal prestige and that of the Royal Gardens.

With Queen Charlotte as their driving force the gardens flourished. At the horticultural helm was William Aiton who continued as superintendent of the amalgamated Kew estates until his death in 1793. He was succeeded by his son, William Townsend Aiton, who later designed the gardens of Buckingham Palace. While the Aitons covered the flowery side of things, the pleasure garden (parts of it by now open to the public two days a week) was enriched by more systematic and adventurous botanical research driven by the forceful figure of Joseph Banks.

Every Blockhead does that

Born in 1743, Banks had the considerable advantage of belonging to the landed gentry and never wanted for money. While many of those who shared his fortunate circumstances might have squandered the lot,

left

William Aiton (1731–1793) by Edmund Bristowe (1787–1876). Gardener to Princess Augusta and George III at Kew from 1759 until his death, Aiton was succeeded by his son, William Townsend Aiton. Here he is portrayed clutching a specimen of Chinese lantern *Aitonia capensis*, a species introduced from the Cape in 1774 by Francis Masson and named in Aiton's honour by Carl Thunberg. The name *Aitonia* was later declared invalid by the botanical powers-that-be and replaced with *Nymania*. The bright red flower on the table is fire heath *Erica cerinthoides*, also from the Cape.

above

The confident and engaging Harrow-Eton-and Cambridge-educated Joseph Banks aged 30, by Sir Joshua Reynolds (1723–1792). An inscription on the letter reads *Cras ingens iterabimus aequor* – 'Tomorrow we sail the vasty deep again.' Banks had become something of a celebrity since returning from the *Endeavour* expedition. By 1773, roughly when this portrait was painted, he had already endeared himself to the king, spent a month at Kew and was establishing 'a kind of superintendence over his Royal Botanic Gardens.'
©*National Portrait Gallery, London*

Banks spent a lifetime indulging his passion for natural history. It goes without saying that this was money well spent.

Similarly, while others of his class and generation were swanning round Europe on the by now obligatory 'Grand Tour', Banks decided he would have none of this. Exclaiming that 'Every blockhead does that', he set off to Labrador and Newfoundland for seven months as naturalist aboard HMS *Niger* in 1766. His appetite suitably whetted, he then paid a cool £10,000 to accompany James Cook on his *Endeavour* expedition to the South Seas to observe the transit of Venus. Banks's assistant was Daniel Solander (1733–1782), a pupil of Linnaeus, and he also took with him three draughtsmen, a secretary, and four servants.

Endeavour set sail on 25 August 1768 and on her three-year voyage visited Madeira, South America, various Pacific islands, New Zealand, Australia (where, impressed with the plants collected by Banks and Solander, Cook named one particular spot 'Botany Bay'), and Java. The ship arrived at Cape Town on 14 March 1771 on the last stop of the return leg of the circumnavigation. By this time the Banks entourage had been reduced to three, the others having died of various illnesses, notably dysentery. In all, 42 of the original crew of 94 had died by the time the ship got back to England, the majority having succumbed to the putrid conditions that prevailed in Java.

The expedition stayed at Cape Town for a month, for half of which Solander was laid up with a severe gastrointestinal condition caught in the Far East. Nevertheless, he and Banks managed to carry out some botanical collecting, as evidenced by the 369 specimens bearing their labels that are now held in the Natural History Museum.

On his return to England in July 1771, Banks began work on cataloguing and conserving his collection, which included over 1,300 new species of plant. Having also become something of a celebrity scientist after the return of *Endeavour*, he and Cook were presented to George III on 10 August 1771. Having successfully manoeuvred his way into royal circles, Banks was appointed 'Scientific Adviser on the Plant Life of the Dependencies of the Crown' in 1772.

Joseph Banks aspired to establish a centre of botanical research at Kew, a facility that he considered would be much more useful than the pleasure garden. He also saw the need for a permanent repository for the living and dried plants arriving from the various colonies and other countries, and for facilities to investigate their economic potential at home and in the all-important empire. The king agreed that this was a worthy goal and Banks became, in all but name until 1797, the first director of Kew.

Cape Town Christened

Back at the Cape, meanwhile, the second half of the eighteenth century was a time of expansion. Its population had reached some 7,000 and in about 1773 the settlement was first referred to, at least by the vising English, as 'Cape Town' as opposed to simply 'De Kaap' or 'The Cape'. By now there was presumably enough infrastructure to merit the urban epithet.

While many of the immigrants chose to stay in and around Cape Town, others headed inland to establish farms or otherwise scratch a living. Observing how the colonists were impacting upon the landscape and its inhabitants, human and wildlife, the botanist Anders Sparrman (1748–1820) visiting in 1775, expressed the view that '…future ages may see this part of Africa entirely changed and different from what it is at present.' Just how 'changed and different' it is today, he would be astonished to see.

Sparrman was one of a significant trio of pioneering botanists who made landfall at Table Bay in 1772. He arrived on 13 April having left his native Sweden in early January. Although only in his early twenties, this was not his first overseas trip as, at the age of 17, he had journeyed to Canton aboard a ship under the command of Captain Ekeberg. On his return from the Far East, Sparrman studied medicine at Uppsala University where:

> his attention was principally engrossed by the science of botany which he pursued with the greatest ardour under its celebrated restorer Linnaeus and became one of his favourite disciples.

A few years later, Linnaeus and Ekeberg persuaded the Dutch authorities to allow Sparrman to visit the Cape as a naturalist.

Having qualified as a doctor, Sparrman, now 24, sailed for the Cape. The voyage was not without incident. Having got only as far as the northern tip of Scotland, the weather took a turn for the worse and 'the wind blowing still stronger … carried away our main-top-sail, though it was quite new, and made of a strong cloth.' Undaunted, and reflecting a youthful exuberance and fascination, he records how 'This ravage and destruction afforded itself a fine spectacle which was to me entirely new.'

Like a Child in a Sweetshop

Once at the Cape, however, Sparrman recognised that his aim to explore and study the natural history of the area could not be done without funds (although he had already secured himself free passage to the Cape). So, over the winter months, he worked as tutor to a family in Simon's Town before moving with his hosts to their summer farm at Alphen further north on the Peninsula. During this time he managed to do some local botanising around the Peninsula and Table Mountain. On some of these excursions he was joined by 'an old Upsal chum, Dr Thunberg … whose taste for botany had induced him to undertake a voyage to this remotest point of Africa.'

Thunberg, professor of botany at Uppsala University, had arrived at the Cape only a few days after Sparrman and was much surprised to see him, considering him at the time to be 'engaged in course of academical lectures' back in Sweden. The two enjoyed the delights of an austral autumn and, despite the wildflowers being at their least floriferous at this season, they were captivated by the botanical riches that this *terra nova* had to offer. Sparrman wrote:

At first almost every day was a rich harvest of the rarest and most beautiful plants; and I had almost said, that at every step we made one or more new discoveries … as I had many Swedish friends, and particularly the great Linnaeus, always present in my memory, every duplicate or triplicate of the plants that I gathered gave me a sensible pleasure.

Like a child in a sweetshop, however, he got a bit greedy:

my covetousness for myself and my friends, frequently induced me to gather more than I was able to attend to and dry in a proper manner. This, doubtless, happens more or less to every botanist who travels into foreign parts.

He lists a great variety of flowers 'strewed over the dry places on the declivity of the mountain' and was particularly impressed with the bulbous species, observing 'Towards spring, divers *ixias, gladioli, moreas, oxalises, mesembryanthemums, antirrhinums* [*Nemesias*] and even various beautiful small irises.' He also mentions osteospermums (Cape daisies), and that 'the *Antholyza* [*Chasmanthe*] *aethiopica* grew from three to six feet in height, with beautiful red flowers.' He notes how the arum lily 'delighted chiefly in moist places near the sea-shore, and was in flower the whole winter.'

Already there is more than an inkling that here was something rather special, and Sparrman recognised this:

Of the plants we met with at this spot, which consisted as well of known plants as those that were quite new, some were rather uncommon, while others were not to be found again in the other places I visited in Africa. Every district has always something peculiar to itself; no wonder then, if Dr Thunberg and I should have passed over various specimens of the vegetable tribe unnoticed, and the common saying *Semper aliquid novi ex Africa,* [Always something new out of Africa; Pliny the Elder] should not still hold good for many years to come.

The fact that there is a high 'turnover' of species over short distances (i.e. the species in one patch of ground may be almost completely different to those found in a similar sized patch even a short distance away), is repeated along the length and breadth of the region and is one of the defining characteristics of the Cape flora.

The move to Alphen gave Sparrman the opportunity to observe and collect as the entourage travelled the few miles north up the Peninsula. He lists a great many plants:

every where by the roadside in their greatest beauty. The pleasure enjoyed by a botanist, who finds all at once so rich a collection of unknown, rare, and beautiful vernal flowers, in so unfrequented part of the world, is easier to be conceived than described.

What a marvellous opportunity this was for someone of his age and interests. Sparrman revelled in all that he encountered, not just the plants. His observations on the Cape's human residents and those *en passant* are frequently insightful, often critical, and occasionally quirky. One of his earliest observations was that the 'Dutch commonly wore

their hats in the house, and that even in company, without it being looked upon as the least breach of politeness.'

He also describes how he and Thunberg:

often enjoyed the company of English ladies, some of whom even staid out our elegant dessert of pipes and tobacco … Some of these ladies came from the East Indies, on their return from Europe, and some from England. The married ones to see their husbands either at Bombay, Madras, or Bengal; the unmarried ones to get husbands … The latter seldom make the voyage in vain, being extremely welcome to such of the single men, as have had time to get a tolerable share of the treasures of India, but could not persuade themselves to wed the dark Indian beauties (as many are accustomed to do) and have not had the leisure to go to Europe merely for the purpose of chusing themselves wives. It was therefore, supposed, that some of the beautiful travellers were actually, in a manner, sent for by commission, though not inserted in the invoice.

At the beginning of September, Sparrman went to visit Thunberg prior to the latter embarking on his first major trip into the Cape interior. Any plans that Sparrman had for similar trips were, however, somewhat delayed when he found himself joining the ship's company of James Cook's *Resolution* which had anchored at Table Bay on 30 October en route to the southern oceans.

Introduced to the ship's naturalists, a German father-and-son combo called Forster, Sparrman recalls that a main topic of their conversation was the existence of *Terra Australis Incognita*. It was thought that such a landmass had to be lurking somewhere down south in order to balance the bulk of land occupying the northern hemisphere, and its discovery was the aim of Cook's expedition. Sparrman commented or, indeed, hinted that:

As the southern continent, which was still pretty generally supposed to exist, had taken no small hold on my imagination, this was sufficient reason for me to congratulate the gentlemen on the trust reposed in them, and the good fortune they had in visiting as naturalists, so distant and unknown part of our globe.

The Forsters took the opportunity to accompany Sparrman on a few botanical excursions round Cape Town. The elder Forster records how these:

always furnished us with an abundant harvest, and gave us the greatest apprehensions that with all our efforts, we alone would be unequal to the task of collecting, describing, drawing, and preserving (all at the same time) such multitudes of species, in countries where every one we gathered would in all probability be a nondescript. It was therefore of the utmost importance, if we meant not to neglect any branch of natural knowledge, to endeavour to find an assistant well qualified to go hand in hand with us in our undertakings.

Cook later related how 'Mr Forster strongly importuned me to take [Sparrman] on board. I at last consented and he embarqued with us

Anders Sparrman (1748–1820)
A talented physician and naturalist and already well travelled, Sparrman enjoyed seven months plant hunting at the Cape in 1772 before joining Cook's second great voyage aboard *Resolution*. He was back at the Cape in March 1775 and spent the next year exploring and collecting.
Wikimedia Commons

Carl Peter Thunberg (1743–1828), the 'Father of Cape Botany'. This painting by Per Krafft the Younger (1777–1863) shows Thunberg at the age of 65 in 1808.

The sash of Commander of the Royal Order of Vaas was added sometime after 1815, the year he was awarded this honour.

In his *Flora Capensis* of 1823, Thunberg lists 2,776 species of flowering plants, 31 ferns, and 17 mosses, representing a third of the species now known to occur at the Cape. This 800-page work represented a monu-

mental step forward in the understanding of the extraordinary botanical richness that exists at the corner of the continent. All in all, Thunberg produced 125 publications on various aspects of the Cape flora.

accordingly…', although the only berth available was 'among the books in the great cabin.'

That Sparrman should have so quickly endeared himself to the Forsters says something for his character, or youthful naivety. The younger Forster seemed a reasonable sort; the elder, in the words of Cook, was:

'an abusive and naturally much abused man … a problem from any angle … from first to last on the voyage, and afterwards, he was an incubus.'

Being difficult and blunt-spoken resulted in Forster being unfairly underrated as a scientist by his contemporaries, despite making signifi-cant contributions to natural history. Ironically, the Forsters had taken the place of Joseph Banks on *Resolution*. Banks had been keen to go but had fallen out with the admiralty over, amongst other issues, their refusal to allow him to be accompanied by a pack of hounds and his personal orchestra.

Sparrman's decision to join Cook's expedition was not as straight-forward as it might appear to us nowadays. Exciting, yes, but potentially very dangerous, with disease and weather causing tremendous hardship and loss of life on sea voyages at the time. So he can be forgiven if he didn't immediately jump at the chance. On the plus side, he was keen that a Swede should 'have the opportunity of making a visit to the south pole, and the continent supposed to be in the vicinity of it' and be party to the 'discoveries of the curious production of nature … in those places.'

On the downside, he pondered:

if my voyage should prove unsuccessful, I was in hopes that my miseries, together with life itself, and all its train of attendant evils, would have a speedy end… Occupied by reflections of this kind I passed the night, perhaps more restless than will easily be imagined. The next morning by day-break, the distraction of my thoughts carried me to my chamber window; here I fixed my eyes on the adjacent meadows, as though I meant to task the plants and flowers that grew on them, whether I ought to part with them so hastily. They had for a long time been almost my only joy, my sole friends and companions; and now it was these only, which in great measure prevented me from making the voyage.

He was clearly quite a philosophical chap, beyond his years, but this does paint quite a touching picture of his insecurity. He then writes:

At length I came to the resolution of undertaking it; yet with a fixed de-termination, that if I had the good fortune to come back to the Cape, I would again occupy myself on this same spot with the most delightful of all employments, the investigation of nature.

Having put his affairs in order, sent consignments of insect and plant specimens to Linnaeus and 'other lovers of the science', and arranged for the rest of his collection to be sent there by his erstwhile employer and landlord 'should he receive any information of our ship's being lost…or give him reason to doubt my return', on 22 November 1772 he was off.

Seven months at the Cape would not, perhaps, have been enough to merit much mention in these pages in terms of his botanical contribution over that short period. But Sparrman's floral guardian angels watched over him and he returned safely to the Cape some three years later to make his mark.

I Joyfully Ran, Sweated and Chilled

By the time Sparrman had slipped over the horizon aboard *Resolution*, the good Dr Thunberg had already been trekking for five weeks or so through the veld (pronounced *felt*; as the natural vegetation or country-side is generally known in South Africa).

Carl Peter Thunberg was born in Jönköping, Sweden, in 1743. He began his studies at Uppsala University in 1761 under the tutelage of Linnaeus, becoming one of his favourite students. After becoming a licentiate of medicine in 1770, he moved to Copenhagen, then to Paris for six months thence to Amsterdam. Here he met up with the Burmans, who enthused him with their collection of plants and botan-ical illustrations from the Cape and the East Indies. The Burmans were later to arrange for him to enter service as assistant ship's surgeon with the VOC.

Thunberg's ultimate aim was to travel to Japan. That country had, since 1635, shut itself off almost entirely from the outside world and the Dutch were the only Europeans currently permitted to enter it. Thunberg was, therefore, instructed to sail with the VOC and break his voyage at the Cape to get a better knowledge of the Dutch language. He conse-quently spent three years there before heading east, visiting Sri Lanka, Java and Japan.

On his return to Sweden in 1779, Thunberg worked under the younger Linnaeus at Uppsala, succeeding to the professorship following the latter's death in 1784. In that year, his first major botanical work, *Flora Japonica*, was published. At the same time he worked on detailed studies of individual plant genera, including *Protea*, *Erica*, and *Restio* from the Cape, producing 25 of these monographs between 1780 and 1821.

Thunberg's first description of the Cape flora as a whole appeared

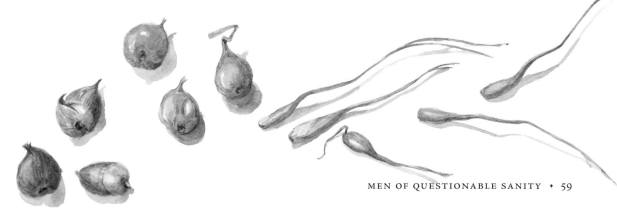

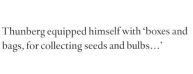

Thunberg equipped himself with 'boxes and bags, for collecting seeds and bulbs…'

in 1794 in the form of *Prodromus plantarum Capensium*, his 'Forerunner of the Plants of the Cape'. A second volume was published in 1800.

With the editorial assistance of Joseph Schultes, this account was later elaborated and published from 1807 onwards under the title *Flora Capensis, sistens plantas Promontorii Bonae Spei Africes: secundum systema sexuale emendatum, redactas ad classes, ordines, genera et species, cum differentiis specificis, synonymis et descriptionibus.*

Known more economically, and affectionately, as *Flora Capensis*, it appeared in 1823 (when Thunberg was 80) in a single, 800-page volume. Representing a monumental step forward in the enumeration and understanding of the Cape flora, it describes 2,776 species of flowering plants, 31 ferns and 17 mosses. As the standard reference for the best part of a century and the solid basis for subsequent work it was Thunberg's greatest contribution to the botany of the region, earning him the accolade of 'The Father of Cape Botany'.

Any lurking doubts that Thunberg had been just a casual traveller and a mere picker of roadside flowers, are dispelled by his own words in the introduction to *Flora Capensis*:

I, Carl Peter Thunberg, a Swede, having been led to these shores … have sought out, carefully and diligently, collected, examined and described, natural objects of all kinds, especially the riches of the Cape flora. With this object I undertook several journeys often fraught with hardships and dangers. Indeed, at first I penetrated every year to the more remote regions, the journeys extending to several months; and thus through sandy *dunes*, treacherous *ravines*, the parched *Karoo*, undulating *plains*, salty *shores*, stony *hills*, lofty *alps*, mountain *precipices*, spiny *scrub*, and rough *woods*, I met the dangers of life; I prudently eluded ferocious tribes and beasts, and for the sake of discovering the beautiful plants of this southern Thule, I joyfully ran, sweated and chilled.

This mildly sensational description might have been better placed within the pages of his popular account, of the sort that would have been sold in the eighteenth century equivalents of airport bookshops, of his experiences at the Cape and elsewhere. Published between 1788 and 1793, Thunberg's *Travels in Europe, Africa & Asia*, or at least the Cape part thereof, doesn't give the impression that he was anything special as a botanist. There are some longer more considered and constructed passages, but in general the narrative bumps along like a fat tourist on a camel. He also felt the need to point out that:

Every traveller thinks himself under an obligation to turn author, and report something marvellous to his countrymen, although, perhaps possessed of so small a stock of knowledge, as not to be able himself clearly to comprehend what he has seen or heard, much less to give others a distinct idea of it. And this circumstance alone has produced more unintelligible books than can easily be imagined.

By this time, some half-dozen accounts of varying length (but generally rather short) had been published about the Cape. 'So much, therefore, having been written, and consequently so much being known concerning this country', grumbles Thunberg, 'I might have saved myself the trouble, and my readers the expense of this publication.' But he didn't.

By the time the reader is in danger of beginning to feel that Thunberg is going to devote as much page space as to why he shouldn't write a book as to the narrative itself, he finally comes to a decision:

But as both my fellow countrymen, and also several foreigners who honour me with their friendship, have frequently signified to me their desire of being informed of the events that have occurred to me, and the discoveries I have made in my travels, and have, moreover, encouraged

It is one thing to be a botanist, a past master in the science, at the head of a botanical department in some wealthy university or well endowed museum, whose patrons are princes, and be able to command the consecutive leisure which goes to the authorship of a great descriptive flora. It is quite another to be a botanical collector, travelling in search of plants through outlandish countries, often in peril of life, always poor and certain to receive little recognition. Yet the number of these labourers who bear the burden and heat of the day, and bring the erudite describer his materials, is by no means small. Nor will they cease from the face of the earth so long as the love of wild nature and a certain strain of gipsyhood combine to make men unable to endure the monotony of labour which brings no ideas.

Professor Peter McOwan, annual address to the South African Philosophical Society, 1886.

Setting off on their collecting trips in the late winter or early spring rains would have given Sparrman and Thunberg the opportunity to see Cape flowers at their best. Thunberg makes rather few references to particular species, but does mention the chincherinchee *Ornithogalum thyrsoides*, noting that 'Tinterintjes is a name given … from the sound it produced when two stalks of it were rubbed together.'

and persuaded me to publish these remarks; I could do no less than (at the few leisure hours I had, after an assiduous application to the duties of my office) collect and put in order the scattered observations I had made in the course of my long-continued and extensive travels.

Thunberg's 'scattered observations' got off to a fairly traumatic start. Having waited a fortnight until fair winds permitted their departure, the fleet of four vessels set off from Texel on 30 December 1771. Less than a week into the voyage the cook somehow contrived to put white lead (an ingredient of paint) instead of flour into the pancakes served up at the evening meal. This resulted in 20 passengers and crew becoming extremely ill. With a bit of impressive detective work, and eliminating seasickness and the lead in the wine cups as a source of the poisoning, Thunberg determined the cause of the affliction, despite himself being subjected to violent and frequent vomiting, excruciating headaches, and a variety of other alarming symptoms. These continued for the best (or worst) part of a month. He survived, although stomach troubles afflicted him for the rest of his life. An account of the course of the poisoning and his treatment of it was published in 1773 and was the first of more than 200 scientific papers that he produced in his lifetime.

In retrospect, Thunberg was doubly lucky. Miraculously, no one died of white-lead poisoning nor of the treatments meted out (including rhubarb and decoction of senna, tobacco enemas, and laudanum and emetics in prodigious quantities), but on a voyage to the Cape that would take 108 days, 115 souls on his own ship *Schoongezicht* died of disease and the total loss on the four ships was 606. The fleet and its depleted complement arrived at Table Bay on 16 April 1772.

The Habits of the Aardvark

Among Thunberg's first observations 'being safely arrived at the Cape of Good Hope' were those on the Company's Garden where he noted that:

> The lime trees do not thrive well, on account of the violent winds that rage here; the same may be said of the hazel, cherry-tree, gooseberry bush and currant-bush, all of which do not thrive and seldom yield any fruit.

He met with the gardener Johan Auge:

> who has made many, and those very long, excursions in to the interior part of the country, and has collected all the plants and insects, which the late governor Tulbagh set to Europe to Linnaeus, and to the Professors Burman and Van Royen. It was of him that Mr Grubb, the director of the bank in Sweden, purchased that fine collection of plants, which was afterwards presented to Professor Bergius, and so well described by the latter gentleman in his book of the Plantae Capensis.

Despite Auge's important contribution, Thunberg seems unable to resist a dig, albeit quickly grudgingly qualified, at the gardener, observing that:

Auge's knowledge of botany was not very considerable, nor did his collections in general extend much further than to the great and the beautiful; but in the mean time, we are almost solely indebted to him for all the discoveries which have been made since the days of Hermannus, Oldelandus, and Hartogius, in this part of Africa.

After a description of the physical and social setting of the settlement, Thunberg's holiday-snaps style of writing quickly engages the reader as he jumps from the habits of the aardvark, through the miscegenation of the colonists' daughters, and on to the price of board and lodging, in successive paragraphs. At the end of August, however, having spent the winter months busying himself around Cape Town, Thunberg at last observed how 'the fields began to be decorated with flowers' and began to prepare for his first expedition.

I therefore provided myself with necessary clothes, as well as with boxes and bags, for collecting bulbs and seeds, a keg of attack [Cape brandy] for preserving serpents and amphibious animals, cotton and boxes for stuffing and keeping birds in, cartridge paper for the drying of plants, tea and biscuits for my own use, and tobacco to distribute among the Hottentots, together with fire arms, and a large quantity of powder, ball and shot of various kinds.

On 17 November 1772, Thunberg obtained the flowers and foliage of Cape chestnut *Calodendrum capense* by blasting into the canopy with his shotgun, a method of collection more traditionally reserved for birds and mammals. This beautiful tree occurs from George on the southern Cape coast north to tropical Africa and grows to 20 metres or more.

The expedition members were Auge ('now to be my sure and faithful guide'), Daniel Immelman the son of an army lieutenant, Christiaan Hardy 'a serjeant, who undertook this tedious journey for the sake of shooting the larger animals and birds', and two Khoekhoen, one to lead and one to drive the covered wagon drawn by six oxen.

It is as well that Thunberg refers to his travelling companions here, because for the remainder of the expedition they get barely a mention and it's almost as if he was travelling alone. A notable exception was his observation that, at one stage in the journey, they wanted to head back to the Cape 'where they might get more wine to drink, and be less frightened by buffaloes.' Thus equipped they set out on 7 September.

Their journey took them from farm to farm up the west coast where:

The sandy low plains, which we traversed, abounded at this time in bulbous plants, beside which others which were now sprung up in consequence of the heavy rains that had fallen during the winter, and which with their infinitely varied flowers decorated these otherwise naked heaths.

The party then turned inland and headed east across the coastal plains to the mountains. A few days were spent resting the draft oxen, providing the opportunity to sort the plants they had already collected and to gather more, including the now famous chincherinchees which he describes thus: 'Tinterintjes is a name given to a species of *Ornithogalum* with a white flower, from the sound it produced, when two stalks of it were rubbed together.'

By early November they had left the mountains and made their way in heavy rain along the less topographically challenging coastal areas towards Mossel Bay. This took them through the valley woodlands and relatively extensive forests that typify the area, contrasting with the almost treeless vegetation inland and to the west.

The 17th of November saw the party returning through Grootvadersbos ('Grandfather's Forest') where Thunberg noted that:

The *Calodendrum* … was then in blossom, the honeyed juice of which I perceived beautiful butterflies sucking, without my being able to reach either one or the other. But by the help of my gun, which I loaded with small shot, and fired among the trees, I got some branches with blossoms on them.

No mention is made of any lepidopteran casualties. The perforated bits of plant, however, provided the material on which Thunberg subsequently based his description of a new species, its name derived from *Kalos*, the Greek for 'beautiful', and *dendrum* [*dendron*], tree. It is popularly known today as Cape chestnut.

Thunberg and his party arrived back in Cape Town on 2 January 1773, having covered some 1,800 km in four months. It would be hard to describe this as a journey of exploration, as they didn't visit any sites that had not already been visited, nor indeed colonised to one extent or another, by Europeans. Exploration or not, there is no doubt that must have been gruelling and exhausting. Any discomfort and deprivation was, however, doubtless compensated for by the wealth of plant material, including many new species, which he had collected.

The Hazard of my Life

Being the height of hot summer in the Cape, Thunberg took advantage of this botanically lean period to organise his specimens and prepare them for shipping back to Europe.

> After having well dried the seeds, spread out the plants, and glued them on imperial paper, packed up the birds and insects, I sent considerable quantities of each to the botanical gardens at Amsterdam and Leyden, by several of the home-ward bound Dutch ships. What I had still remaining, I divided into different parcels, and packed up for my patrons and friends in Sweden, especially the Archiaters [chief physician to the king] and chevaliers [knights] von Linné and Back, Professor Bergius and Dr Montin; these I had an opportunity of sending in Swedish ships by favour of several gentleman officers who honoured me with their friendship.

Thunberg spent the following months making short excursions round Cape Town. He climbed Table Mountain 15 times and was particularly impressed with its orchids, notably red disa *Disa uniflora*, and the delightful pale-blue drip disa *D. longicornu*. The latter, he relates, were obtained with:

> great difficulty, and at the hazard of my life … This plant grew in one spot only, on a steep rock, and so high up, that in order to come at it after we had clambered up the side of the rock as high as we could, I was obliged to get upon the shoulders of M Sonnerat [his French companion], when, with a long stick, I beat down five of these plants, the only specimens that were then in bloom.

M Sonnerat, it was noted, made a collection of 300 species of plant on that one excursion and, having worn out three pairs of shoes, descended the mountain barefoot.

He also visited Robben Island in Table Bay, describing it as:

> formerly the resort of a great number of seals...but now these animals having been driven away from it, it is become the retreat of chameleons, quails and prisoners for life.

An interesting excursion was made in March to the southern parts of the Cape Peninsula. Thunberg apparently didn't get as far as the Cape of Good Hope in its strictest sense, which is a little surpriing given that the nature of the landscape and the difference in topography might have held new plants. Apart from anything else, one would have expected him to be curious to see that '… great and famous cape concealed for so many centuries' and so celebrated by historians and navigators.

This particular trip was significant, however, in that he was accompanied by 'Mr Gordon and an English gardener, lately arrived, of the name of Masson.'

With temperatures rising at the onset of summer, most of the bulbs quickly go to seed and their foliage withers. At the hottest time of year in late summer and early autumn, however, the leafless stalks of *Haemanthus coccineus* emerge from the dusty soil and come into full bloom. This species was already familiar to European botanists through the efforts of the early, and largely opportunistic, collectors of the seventeenth century, and was one of a number of plants sold under the name 'Cape tulip'.

CHAPTER FIVE

The Scots Garden-Hand

The 'English gardener, lately arrived' was, in fact, a Scotsman, Francis Masson. Born in Aberdeen, in August 1741, nothing is known of his early life, although he clearly received some formal schooling. As one authority, Frank Bradlow, observes from a study of his journals, 'Masson's literary style ... is certainly not that of a man lacking in education which, after all, was a Scottish one – probably the best in the world at that time.' Masson must also have trained as a gardener, as at some point he gravitated south to London and found work at Kew where a number of Scots were already employed. (Joseph Banks ascribed the eagerness to employ Scots to his estimation that 'so well does the serious mind of a Scottish education fit Scotsmen to the habits of industry, attention and frugality, that they rarely abandon them at any time in life, and I may say, never while they are young.')

Another of these 'Scotch Kewensians', as they were labelled, was William Aiton who, as superintendent of Kew, was Masson's immediate boss under the overall direction of Banks.

In order to realise his ambition for a centre of botanical and horticultural excellence at Kew, Banks was determined to obtain plants from as many parts of the world as possible. His diary account of the month he spent at the Cape does not give much indication that the location and its plants had impressed him that much, although concern for Solander's health had been a time-consuming distraction. Perhaps

he did recognise that in its extraordinary botanical richness lay the potential for great scientific gain but, for whatever reason (and it could have been as simple as the fact that there was a particular ship going that way), the Cape was chosen as the destination of Kew's first dedicated international plant collector.

While it was almost certainly Aiton who suggested him for the post, an official approach to engage Masson was apparently made to the King on behalf of Banks by Sir John Pringle, president of the Royal Society. Perhaps 'habits of industry' had less to do with Masson's appointment than chauvinism on the part of Aiton but his choice was, in due course, certainly vindicated.

Masson's appointment represents a major landmark in the history not just of plant collecting in the Cape but in the development and expansion of Kew. Masson simply observed, with a touch of wry self-effacement, that 'his Majesty was graciously pleased to adopt the plan, though at that time so little approved by the public, that no one but myself chose to undertake the execution of it.' Nevertheless, he was to receive recompense of £100 a year (payable on his return) and an expense allowance of £200 per annum. As Captain Cook earned less than £200 a year, this wasn't bad going.

Masson was dispatched to the Cape aboard Cook's *Resolution*, leaving Sheerness dockyard on 21 June 1772. Joined by *Adventure* at Plymouth,

---

Scotch gardeners are often preferred in England, as if they had a better knowledge of their profession than English gardeners. As gardening is certainly as well understood in England as it is in Scotland, it may be worth while to enquire into the cause of this preference; and also the reason why so few young men in England, after serving a regular apprenticeship as gardeners ever arrive at the head of their profession. They remain only a step above a common labourer, and seldom remove from the place of their birth; while most young men who learn gardening in Scotland become in time head gardeners, either at home or abroad. Various reasons are assigned why preference is given in England to Scotch gardeners: one is, that they are usually better educated; another, that the greater coldness and changeableness of the climate in Scotland obliges the gardener to take greater care and pay more attention, which renders him more skilful in his business; another case assigned is, that Scotchmen are generally a more steady and calculating race.

John Wighton, gardener to the Earl of Stafford; *The Gardener's Magazine*, 1840

the ships sailed via Madeira and Cape Verde, dropping anchor in Table Bay on 30 October.

What Masson felt about being deposited at the Cape while the rest of the ships continued on their epic voyage a few weeks later (with Sparrman now aboard), is not recorded, but he was probably not too upset to part with the Forsters. Of Masson, the elder Forster had little more to say than that he was 'a Scots garden-hand.' Granted, Masson was no intellectual, and had neither the academic mind nor qualifications of the scientists who were his contemporaries, but this does not detract from his achievements over two sojourns that eventually amounted to 12 years in South Africa. In fairness, it should also be pointed out that in a letter to Banks of 30 January 1786, Forster refers to Masson as one 'whose worth and excellence there is no need of bearing testimony to you'.

Masson was to remain at the Cape for two-and-a-half years on this, the first of his visits. During this time he made three journeys into the interior, a short narrative of which was published in the *Philosophical Transactions* of the Royal Society in 1776 under the title '*An Account of Three Journeys from the* Cape Town *into the Southern Parts of* Africa; *undertaken for the Discovery of new Plants, towards the Improvement of the Royal Botanical Gardens at Kew.*'

Masson's '*Account*' was additionally significant in being the first personal record of travels in South Africa to be published in English. There is no indication of how it was received by the scientific community, but a critique by the elder Forster was true to character and less than flattering:

> I have yet to come across a translation of this poor product, and if all our publishers, who are only too eager to print almost anything, really had projected it, then this would fully establish its value for our public, so we need not waste words over it.

Whatever Forster's crude assessment may have been, the significance for us is that Masson had become the first to describe in print, in some detail, the unique vegetation of the southern and south-western Cape, which was in effect the Cape Floral Kingdom. Although the phrase was not coined until 1908 (by the distinguished, German-born botanist Rudolf Marloth, 1855–1931) to describe what is the smallest and, for its size, richest of the world's floral kingdoms, it was the pioneering efforts of Masson and his contemporaries that began the long process of the Cape's recognition as a remarkable region for plants, with an all but unparalleled botanical diversity and richness.

Many of Masson's travels were through the Cape Floral Kingdom's most extensive and diverse vegetation, a shrubby heathland called fynbos (pronounced *fayn-boss*). This is characterised by a particular abundance of plant species, the most distinctive and best known of which are the ericas and the proteas.

Francis Masson (1741–1805), the 'Scots garden hand' who became the first dedicated plant hunter sent overseas by Kew. Writing in 1920, James Britten of the Kew Herbarium asserts that this portrait is 'that which was at one time in the possession of James Lee; it was bought from a general dealer at Hounslow in 1884 by Mr. Carruthers and was by him presented to the Linnean Society.

Masson is shown with a background of cliffs: on the left is a view of Table Bay viewed from the north; Devil's Peak, Table Mountain, and Signal Hill are readily distinguishable; the artist is unknown.' The artist is now recognised as George Garrard (1760–1826), and the work still hangs in the Linnean Society in Piccadilly.

By permission of the Linnean Society, London

These Sequestered and Unfrequented Woods

Masson set out on his first plant-hunting expedition, a comparatively short journey in distance and duration, on the evening of 10 December, 1772. His party comprised Franz Pehr Oldenburg, a Swedish mercenary who had served in the VOC, and a Khoekhoen who drove a wagon hauled by eight oxen. Oxen were preferred to horses because, according to Masson, 'they are much cheaper, and more easily maintained'. They also browsed the heathy scrub, to which horses were averse.

Their journey took them north-east out of Cape Town along the edge of the wide sandy expanse of the Cape Flats to the low hills of Tygerberg and on through the gentle, shaley downs to Paardeberg. Tygerberg translates as 'Leopard Hill' and was so called either because it was the haunt of Leopards or because its slopes are dappled with spots, the consequence of countless generations of termite activity; both explanations have their merits and adherents. Paardeberg is 'Horse Mountain', although the 'horses' were actually zebras.

Appreciating that a good deal of the time was spent in collecting plants, it took Masson's party six days to travel a distance that today could be driven in less than an hour. They then took something of a circuitous route around the mountain valleys of Paarl, Franschhoek and Stellenbosch before heading over the Hottentots' Holland range and east along the base of the Cape Fold mountains that rise steeply from the southern Cape coastal plain. Of the Hottentots' Holland, Masson observed: 'These mountains abound with a great number of curious plants and are, I believe, the richest mountains in Africa for a botanist.' This was an astute observation as the heart of the mountains, the Kogelberg, supports almost 2,000 plant species. (Kogelberg translates as 'Cannonball Mountain' and may refer to the far-carrying rumble of the stones on the beaches grating and grinding together in the receding waves.)

Much of their route traversed difficult terrain and Masson admits that he found the going tough. On 10 January 1773, while exploring the forests in the eastern part of the Riviersonderend (river without end) Mountains, he made one of the few references to the physical demands of his travels, writing that he 'endured this day much fatigue in these sequestered and unfrequented woods, with a mixture of horror and admiration.'

His outward journey came to an abrupt halt on 18 January. By this time he had reached the Breede (Broad) River at Swellendam, almost 200 km east of Cape Town, and 'finding the season too far spent for making any considerable collections, I returned back to the Cape by the same road I came.'

Although there is no journal record, it can safely be assumed that Masson spent the autumn and winter months sorting his collections and making short trips of a day or two in the immediate vicinity of Cape Town, an extremely profitable collecting ground for the more sedentary botanist. What is known, however, is that Masson accompanied Thunberg and Robert Gordon of the Cape garrison, on a week-long tour of the Peninsula mountains in May 1773.

Like so many scarlet Lilies

Thunberg himself undertook a number of short trips over the winter but, by springtime, was eagerly anticipating another major journey into the hinterland. 'The month of September was already begun,' he

Robert Jacob Gordon (1743–1795; artist unknown) was born in Holland where his father commanded a company of the Scots Brigade into which Robert was commissioned. 'His mind thirsting for a variety of knowledge', Gordon went to the Cape in 1773, spending much of that year exploring the interior. He returned to Holland in 1774 but secured a posting back to the Cape in 1777. In 1780 Gordon was appointed garrison commandant. Over the next 15 years he undertook many important expeditions, recording his discoveries in detailed journals, maps and watercolours. In 1795 the Cape surrendered to the English. Torn between loyalty to the Prince of Orange and duty to the Dutch Republic, Gordon was perceived as being irresolute during the invasion. Publicly insulted by his own men, and likely already ill, he committed suicide on 25 October. *Image courtesy of Iziko Museums of Cape Town*

enthused, 'and the beautiful and flowery spring making its appearance, put me in mind of preparing for a long journey up the country.'

A major problem, however, was that he had run out of money or, as he put it, 'The trifling viaticum I had brought with me from Europe, I had long ago consumed.' To make matters worse, his official post as surgeon with the VOC had not been confirmed because the muster-roll had been left behind in Europe and no one could be paid until it was received. This resulted in the ship's company also not receiving payment nor being able to go on leave for two or three years.

Unwilling to remain 'an idle spectator at the Cape' Thunberg fell back on the understanding of the secretary of the Council of Policy, Mr Bergh:

> who had not only hitherto kindly assisted me with his purse, but also generously opened it to me on this occasion, and thereby enabled me to make another excursion into the interior part of the southernmost point of Africa.

He was disinclined to call upon the services of the sergeant and gardener who 'the year before greatly contributed to render a small cart still more insufficient for my wants.' This time he acquired a new wagon and loaded it up with the requirements for the journey, including:

> several medicines to distribute among the colonists in the interior parts of the country, who might stand in need of them, and had before on occasions shown me the greatest kindness.

Thunberg's travelling companion was Masson and together they:

> formed a society, consisting of three Europeans and four Hottentots, who for the space of several months were to penetrate into the country together, put up with whatever we should find, whether good or bad, and frequently seclude ourselves from almost all the rest of the world, and of the human race.

Last summer I duly received by your incomparable kindness two packets with so many rare plants that it made me dizzy; I took the greatest delight in them from the mid-summer. For every plant I am most obliged to you Hr. Doctor. God grant that I may live to see the day when I may talk with you about it.

Linnaeus, letter to Thunberg, 29 October 1773.

You will greatly satisfy the directors of the Hortus [Amsterdam botanical garden] if you would send to the Hortus a box filled with living plants, and would select among other things those plants which can be propagated by seeds, like geranium spinosum, flavum, Mesembryanthema, Cactus, Aloes and similar plants, especially Succulents which are particularly required by the herbarium and the hortus in question, and which must be put under the charge of the Captain, by you as well as the governors of the Cape of Good Hope.

Kindly add a number of bulbous plants and those species of geranium which are distinguished by their turnip-like roots, and some seeds selected from the rarer specimens.

Nicolaas Burman (1733–1793), professor of botany at Amsterdam, letter to Carl Thunberg, 30 August 1773.

Among the 'great number of beautiful plants, particularly ixiae, irides, gladioli' about which Masson enthused, would have been *Ixia paniculata*. One of 50 species of Ixia (all elegant to a degree) in the Cape, it grows on sandy slopes and flats, flowering from October to December.

The third European was Masson's driver (they had a wagon of their own) and the party set out on 11 September 1773.

Their course took them, firstly, along the coastal tracts north of Cape Town up the west coast through the Swartland, a level or gently undulating landscape of fertile shales. Thunberg noted, in one of very few insights into his plant collecting techniques:

> It is only in the spring and in the beginning of summer, that these low sandy plains are adorned with flowers … After the south-east winds and the droughts have set in, the seeds of these flowers are quickly scattered over the fields, often before they are quite ripe. For this reason I was obliged, when making collections for the botanical gardens of Europe, especially of the annual plants, to gather the seeds in an unripe state, and lay them up in paper to dry and ripen gradually.

As they journeyed north, Masson was captivated by the flowers that he saw, as is any visitor today who visits the area in spring. He certainly waxes quite lyrical: 'The whole country affords a fine field for botany, being enamelled with the greatest number of flowers ever saw, of exquisite fragrance and beauty.' He also describes the waterside birds, in particular the gaudy red bishops (a species of weaver-bird) which 'when sitting among the reeds look like so many scarlet lilies'. On reaching the Berg River he collected 'a great number of beautiful plants, particularly *ixiae, irides, gladioli.*'

Across the river the party headed through the Kardouw Pass and experienced increasing difficulties as the terrain became rougher and there was incessant rain. They reached the other side after skirting 'precipes...so steep that we were often afraid to turn our eyes to either side.' They lodged overnight in a Dutchman's one-roomed hut 'where he and his wife also slept; and in the other end lay a number of Hottentots promiscuously together.'

After four days travelling along the northern bank of the Olifants (Elephants) River, their route was blocked by a ridge of mountains and the wagon overturned and was damaged in attempting to surmount them. This is the Koubokkevelde, the 'Cold Country of Antelopes' where Masson saw the Springbok after which it is named. He wouldn't see one now.

Heading south-east through Elands Kloof, the landscape continued to hold Masson in some awe:

> The road is rugged beyond description, consisting of broken and flattened rocks and rugged precipices, encompassed on each side with horrid impassable mountains; the sides of which are covered with fragments of rock that have tumbled down from the summits at different times.

Later on, he writes that:

> The country is encompassed on all sides with very high mountains, almost perpendicular, consisting of bare rocks, without the least appearance of vegetation; and upon the whole, has a most melancholy effect on the mind.

Masson and Thunberg then sent their wagons and servants on a gentle route to Tulbagh. They themselves opted to go through the mountains via a particular pass, despite being warned that 'without a guide we should run the risk of losing our lives.' Masson admitted that:

> We were a little intimidated by this information, but fortifying ourselves with resolution we proceeded and in an hour arrived at the first precipice, where we looked down with horror on the river, which formed several cataracts inconceivably wild and romantic.

There was certainly something of the poet in Masson, albeit largely suppressed and self-conscious.

It took them three hours to negotiate the pass, its precipices and fords, but they thought their:

> labour and difficulties largely repaid by the number of rare plants we found here. The bank of the river is covered with a great variety of evergreen trees...and the precipices are ornamented with *ericae* and many other mountain plants never described before.

After a day's rest, they mustered the energy to climb the Groot Winterhoek and, with the ultimate destination of his plant material in mind, Masson hoped to find plants on its summit 'that might endure the severity of our climate.' This was a not unreasonable expectation, as the mountain is more than 2,000 m high and its summit receives a fair amount of snow in the winter. They were, however, disappointed to find only a few grasses and Cape reeds.

On 28 October the party discovered an enigmatic green-flowered *Ixia*. Thunberg described it in his account as 'a green variety of the *Ixia maculata*, another tall bulbous plant, which is as elegant as it is singular, with its long cluster of green flowers growing out like an ear of corn, and is so extremely scarce all over the world.'

In his account of that day, Masson describes collecting:

> many remarkably fine flowers, particularly one of the liliaceous kind, with a long spike of pendulous flowers of a greenish azure colour, which among the long grass had an admirable effect (this is *ixia viridis*).

This beautiful plant is now recognised as a species in its own right, *Ixia viridiflora,* the green ixia, and it grows only in and around the Tulbagh Valley, an area renowned for its natural beauty and remarkable diversity of plants.

In this case, Masson was nearer the mark than Thunberg, although differences in plant identification are not the only discrepancies to arise between their respective accounts of their journey. At one point, for example, Masson describes how at a flooded river:

> The doctor imprudently took the ford without the least inquiry; when on a sudden, he and his horse plunged over the head and ears into a pit, that had been made by the hippopotamus amphibious, which formerly inhabited those rivers. The pit was very deep, and steep on all sides, which made my companion's fate uncertain for a few minutes, but, after several strong exertions, the horse gained the opposite side with his rider.

Thunberg having, he claimed, been given dodgy directions by a local Khoekhoen, took the plunge and describes what happened next.

The original specimen of groenkalossie or green ixia *Ixia viridiflora*, in Thunberg's herbarium at the University of Uppsala. He described it as 'another tall bulbous plant, as elegant as it is singular' when he picked it at the Cape on 28 October 1773.
Botany Section, Museum of Evolution, University of Uppsala

> I, who was the most courageous of any of the company, and, in the whole course of the journey was constantly obliged to go on before and head them, now also, without a moment's consideration, rode plump in to the river, till, in a moment, I sank with my horse in to a large and deep sea-cow hole, up to my ears. This would undoubtedly have proved my grave if my horse had not by good luck been able to swim.

He goes on to explain how he was able to 'possess himself in the greatest dangers', and 'with the greatest calmness and composure guided the animal and kept fast in my saddle.' He finishes his heroic account by describing how 'All this time my fellow travellers stood frightened on the opposite bank and astonished, without daring to trust themselves to an element that appeared to them so full of danger.'

Masson's account of this incident appeared in print some years before Thunberg's, so maybe this is the latter's attempt to give a favourable version of events.

By the time they reached the Gourits River in mid-November, the expedition was some 300 km east of Cape Town. Here Thunberg observed how:

> The whole of this tract produced aloe bushes in abundance, which in some places entirely covered the hills and the sides of the mountains, where they appeared at a distance like a numerous army.

Sap from the aloes (in this case of *Aloe ferox* – the 'fierce aloe', on account of its spines) the theraputic virtues of which are well known, was harvested by a farmer called Peter de Wett who, as the first local to do so commercially, 'was said to have the exclusive privilege of delivering and selling it at a certain price to the Company'. The extracted juice was boiled down in 'English iron pots' to a thick consistency, then left in boxes to set into solid gum weighing between 150 and 250 kg.

After five days journeying through the Little Karoo, where they encountered:

Groenkalossie *Ixia viridiflora* is becoming increasingly rare in the wild, but its enigmatic appearance and relative ease of cultivation should at least ensure its survival in 'captivity'.

an infinite number of evergreen shrubs, both frutescent and succulent: among the latter we found many new species of *crassula, cotyledon, euphorbia, portulaca, mesembryanthemum,*

the botanists then 'resolved to visit the seashore' and reached the south coast at Mossel Bay on 16 November. After a few days fossicking among 'shrubs of various kinds; the greatest part of which were unknown to us, and many we did not find in flower' they headed inland once more, striking north-east over the Attaquas Pass, named after a chief of the local inhabitants, the Hessequa Khoekhoen. Once over the pass they:

continued [their] journey through a dismal valley, where we saw neither man nor beast; but our labour was generously rewarded by the productions of the vegetable kingdom, having found several new species of plants, which for neatness and elegance exceeded any thing I had ever seen.

Masson never lost his enthusiasm for new discoveries and for the beauty of the plants he encountered; these were, indeed, the reward for the painful slog up and down the mountains and across dry, dusty plains, one of which they encountered the next day.

At night we got clear of the mountains, but entered a rugged country, which the new inhabitants name Canaan's Land; though it might rather be called the Land of Sorrows; for no land could exhibit a more wasteful prospect; the plains consisting of nothing but rotten rock. Yet notwithstanding the disagreeable aspect of this tract, we enriched our collection by a variety of succulent plants, which we had never seen before, and which appeared to us like a new creation.

An eight-day stop-off at 'the house of Jacob Kock, an old German, who used us with great civility' allowed Thunberg and Masson to recharge their batteries and to do some serious botanising in the vicinity of the Seekoei (Hippopotamus) River at Humansdorp.

We found here a new palm. We observed two species; one about a foot and a half in diameter in the stem, and about twelve feet high … The other sort had no stem, with the leaves a little serrated, and lying flat on the ground, which produced a large conical frutification about eighteen inches long and a foot or more in circumference…

These were not palms, but cycads, and we know that on 1 December 1773 Masson collected at least one specimen of the first one that he describes. This was *Encephalartos altensteinii,* and the young plant that he procured and sent alive to Kew still survives.

What Masson charitably does not mention is that Thunberg was confined to quarters having been badly sunburned. It was left to the latter to describe how, in the shallows of the river, he had:

walked about for several hours quite naked … collecting insects and shrubs … with nothing but a handkerchief about my waist, not suspecting that the sun beams would have any bad effect upon me. But, in a short time, I found that all parts of my body which was above the water, was red and inflamed. This disorder increased to such a degree that I was obliged to keep my bed for several days, and could not even bear a fine

calico shirt on my body … till I had anointed myself with cream [of the dairy variety] in order to lubricate my parched skin.

It is, perhaps, surprising that sunburn was not a more common affliction of travellers, especially near the sea when cool winds belie the fierceness of the sun.

Masson and Thunberg arrived at the Sundays River having travelled through:

a pleasant country, diversified with smooth green hills, interspersed with evergreens … which together with the fine disposition of the woods and groves, could not but charm us, who, for upwards of three months, had been climbing rugged mountains, and crossing sultry desarts.

This was to be the limit of their journey as their local guide and native servants 'refused to advance further' because of the hostile tribes beyond. Masson noted:

In consequence of their remonstrances, and the bad state of our carriages … being ready to drop to pieces, and many of our oxen sick, we, with much reluctance, consented to return the same way they came.

A field sketch of the inflorescence and individual flowers of Guernsey lily *Nerine sarniensis.*

While Sparrman and Thunberg's main interest lay in collecting for taxonomic study, Masson focussed more strongly, if not exclusively, on collecting plants for cultivating. Among those he sent back were Guernsey lily *Nerine sarniensis* (centre) and a range of other bulbs from which commercial varieties were later developed. Clockwise from the bottom left are the cultivars *Gladiolus* 'Frizzled Lace', *Tritonia* 'Serendipity', *Tritonia* 'Bridal Veil', and *Gladiolus* 'Gwendolyn'.

The homeward journey took about five weeks but is reduced in Masson's account to a page or two in his journal; obviously, the travellers were by this time weary and keen to get back to their base in Cape Town. They made a brief diversion into the Karoo, where they got lost, and there are passing references to the heat, lack of water, sickness of the oxen (they suffered from hoof rot), and collecting the occasional plant, including species of *Erica*, *Pelargonium* and *Stapelia*. Otherwise, the return trip appeared relatively uneventful and they arrived in Cape Town on 29 January 1774. The expedition had taken them 4 months and 14 days, and they had covered some 2,300 km.

Throughout their travels, Thunberg or Masson rarely had to camp in the open, which would be the expected way to spend the night in such a remote and sparsely populated region. At the end of each day, however, they almost invariably fell upon the hospitality of the farmers whose homesteads were scattered throughout the region. Very rarely were they turned away, despite the fact that many of the farmers lived in very humble circumstances. This demonstrates, firstly, the hospitality of the settlers and, secondly, just how extensively they had spread, especially along the southern Cape coast, within not much more than 100 years of settlement.

Being back in the relative comfort of Cape Town was no excuse for not continuing to collect plants, although neither botanist is very forthcoming about what they got up to in the bright lights of the city, or what passed for them at that time. At one point, however, Thunberg describes how 'At my leisure hours I never neglected to visit the hills, mountains, and fields, near the town.' He also mentions accompanying one Lady Monson on expeditions to farms. In case there is any suggestion of ulterior motives for their floral dalliances, it should be mentioned that Lady Monson, a great-grand-daughter of Charles II, was aged 60 and married to a colonel of the regiment in the East Indies. Oh, and Masson went too. Lady Monson was also a respectable plantswoman in her own right, a talent recognised by Linnaeus who named the genus *Monsonia*, in her honour.

Masson also has little to say about what he did over the winter, but it is known that he established a garden in which to keep his plants and presumably he busied himself with tending them and sorting out seeds, bulbs and live plants to send back to Kew.

Enamelled with Flowers

After a damp urban winter, Masson and Thunberg were keen to set out again as the days warmed and the rainfall slackened, and the two prepared for a journey that would take them further north than their previous expeditions. Thunberg started out on 29 September 1774, on what was to be the last of his Cape collecting trips.

Masson set off a few days earlier than Thunberg but managed only a short distance on account of the 'badness of the weather'. The clouds then cleared and, crossing the Cape Flats almost to Stellenbosch, he found the 'whole country enamelled with flowers'.

The two met up at Paarl at the beginning of October. Their subsequent journey can be described as a northern loop, taking them to the northern limit of the Cape Floral Kingdom at Nieuwoudtville, then out into the Roggeveld Mountains of the Karoo eastwards as far as Sutherland, thence back down through the mountain passes to Cape Town.

By 6 October they had progressed northwards through the west coast hinterland as far as Paardeberg, on whose summit 'we found a treasure of new plants, which we had not seen before.'

These granite outcrops contain different species from the low-lying areas of shaley-soiled renosterveld of the valleys below. Renosterveld means 'rhinoceros bush' and is the second major vegetation type, after fynbos, of the Cape Floral Kingdom. Even in Masson's day, it had already been identified as fertile and productive and was undergoing rapid transformation to agriculture, with the concomitant and precipitous loss of its black rhinos as well as a host of plants, particularly bulbous species. Masson notes how this country 'has but a barren appearance; yet contains several rich plantations, producing abundance of corn and wine: and the peasants live luxuriously.'

Further north at a mountain outcrop called Riebeek-Kasteel ('van Riebeek's Castle', named in 1661) they 'collected here many remarkable new plants, in particular a hyacinth, with flowers of a pale gold colour.' This is most likely *Lachenalia aloides* but Masson volunteers no more than the fact that they discovered and collected it. Thunberg, in contrast, uses the opportunity for self-congratulation to best effect:

> whilst we were searching here after some curios plants, and laying them up in our books, I stumbled upon a very near, but, at the same time, dangerous way, to get to the other side of the mountain's perpendicular flanks. This was chink a few fathoms length, and so narrow as to be capable of admitting a middle sized man only. Through this I ventured to crawl on my hands and feet, and was fortunate enough to get safe over to the other side, whence it was only the distance of a musquet-shot to

The collection of this *Lachenalia aloides* is described in heroic detail in Thunberg's journal. It is a highly variable species, coming in a variety of patterns of red, orange, yellow or greenish-blue.
Botany Section, Museum of Evolution, University of Uppsala

A widespread species of orchid, occurring from the western Cape north to Zimbabwe, *Disa cornuta* would have impressed Masson and his fellow plant hunters, not least because it can grow up to a metre in height. These relatively short specimens are typical of those found in the windswept tracts of the southern Cape Peninsula.

our conveyance. My fellow traveller, together with his dog, stood astonished at my adventurous exploit, the one howling, the other almost crying; and at the same time vexed to think that he should be obliged to go alone a long way round about, without once daring to take the direct path. My courage was rewarded with a small plant which I got in the chink, and which afterwards I sought in vain in other places.

By 25 October they had reached the Olifants River where 'The sterile appearance of this country exceeds all imagination; wherever one casts his eyes, he sees nothing but naked hills, without a blade of grass …' Masson was surprised, therefore, to learn from a local farmer how 'in winter the hills were painted with all kinds of colours.' These are the annuals and succulents that provide such a brilliant patchwork in springtime if there have been ample winter rains. His informant went on to say that 'it grieved him often, that no person with a knowledge in botany had ever had an opportunity of seeing his country in the flowery season.'

The following part of their journey took them 300 km north of Cape Town and out of the Cape Floral Kingdom into the Karoo, entering the former briefly again only when they came to Nieuwoudtville. The desert was inhospitable and they:

suffered from the heat of the Sun and want of water, but our sufferings were still aggravated when we thought on our poor animals, who often lay down in the yoke during the heat of the day…We were obliged to make the greatest expedition to save the lives of our cattle, only collecting what we found growing along the roadside, which amounted to above 100 plants, never before described.

The fickleness of the climate later manifested itself when:

it blew a violent storm and was extremely cold. The next morning the ground was white with frost, and there was ice upon the pools as thick as a crown. This alarmed the peasants.

The party arrived at Nieuwoudtville on 4 November. By this time, the spring flowers would have been well over. Although the masses of seed-heads and bulbs with their associated shrivelled leaves would have given them some idea of the floral riches, it is a pity that these early botanists did not witness the glory of the spring wildflower display here. In a good year the flower–strewn plains and rocky outcrops of Nieuwoudtville represent, quite simply, one of the most extraordinary and breathtaking spectacles in the natural world.

Nieuwoudtville stands near the northern limit of Cape Floral Kingdom. A long, finger-like projection of the Bokkevelde Mountains, replete with fynbos and renosterveld, extends northwards beyond the town for a few kilometres before being lost in the rocky plains of the Karoo. If there was any doubt that they had left behind the distinctive Cape vegetation once they headed into the Karoo, it is dispelled by Masson's observation that 'I did not see an erica or a protea in the whole country'. He also comments that they 'had excellent bread, good mutton, butter and milk, but no kind of strong liquors.' It is difficult to tell if the last is noted for regret or approval.

The entrance back into the Cape Floral Kingdom is at the famous Karoopoort, literally 'gateway' or 'entrance' to the Karoo. Here, on the route from Ceres to Calvinia, with an abrupt change in soil type and underlying geology, the Karoo gives way to fynbos over a sharp ecological divide the width of a dusty road. Masson and company passed through this portal on 8 December.

About eleven o'clock at night we got clear of the desert, and arrived at the foot of the Bocke Veld mountains, where we lodged by a rivulet of pure fresh water; and we spent the remainder of that night and part of the next day in great luxury.

This contrasts with recent days when they had found water to be a rare commodity or virtually unpotable, as he describes at one spot:

The small quantity of water which was here to be found in a very few places and in small cavities, was not only salt, but likewise so thick and turbid with clay and other impurities, that we were obliged to lay a handkerchief over it, in order to suck a little of it into our mouths.

As with the return leg of their previous journey, Masson and Thunberg seem to have got the bit between their teeth and sped back to Cape Town. It took three weeks but, in Masson's case, only one journal page to traverse the mountains, and Thunberg doesn't have much more to say about it. They were back at the end of December, having travelled about 900 km in just under three months.

This marked the end of a great plant-hunting partnership. Although they didn't have much to say about each other, and both had the occasional snipe in their respective accounts of their travels, there could be no doubt that Masson and Thunberg enjoyed mutual respect and affection. Such travels as they undertook would have been very difficult otherwise. Later correspondence between the two is certainly warm and civil, without ever being effusive (although Masson does address Thunberg as 'Much esteemed friend'). Scots and, I daresay, Swedes, are not renowned for overt displays of affection. If it's a weakness, blame Calvinism.

Masson's Eponymous Plant

Thunberg sailed on 2 March 1775 from Cape Town for Batavia, thence to Japan. He stopped off at the Cape in early 1778 on his way back to Europe, calling in at Amsterdam, then London where he visited Banks, and James Lee's Vineyard nursery at Hammersmith, already famous for its range of Cape plants. He finally returned to Sweden in March 1779.

A few weeks after Thunberg's departure from the Cape, Masson also took his leave and sailed back to England. This was no holiday, however, and he set about putting his collection in order and writing an account of his sojourn at the Cape for publication.

He wrote to Linnaeus on 26 December 1775, sending with his letter a dried plant specimen.

The inclosed specimen I think is a new genus, to which my worthy friend Mr. Thunberg had a great desire of giving the name *Massonia*, honouring me with this mark of friendship. But notwithstanding the good will of Dr. Thunberg and many other botanical friends, I have declined receiving that honour from any other authority than the great Linnaeus, whom I look upon as the father of botany and natural history, in hopes that you will give it your sanction. I am sorry that the leaves are not more perfect; but it is the only specimen I have.

Masson was granted his wish. In a letter of 6 August 1776 he wrote 'Your good nature and generosity, in acquiescing to confer the honour of a genus on so young a botanist as I am, deserve my most grateful thanks.' Subsequently, however, the authority for naming *Massonia* reverted to Thunberg because, as the first to name it, he remains the valid authority under the rules of taxonomic nomenclature.

Locust Soup

A few days before Masson left for England in March 1775, *Resolution* hove to in Table Bay on the last leg of Cook's voyage. Anders Sparrman was back in town and ready to take up where he had left off after three years at sea. This involved, firstly, earning some money to finance a collecting trip. This he did by working as a physician and by selling his Swedish-to-English translation of a treatise on the diseases of children. Within three months he had raised the required funds and made the preparations for an expedition to the eastern Cape.

'Nature has presented herself to me in various shapes, always worthy of admiration, often enchanting, and sometimes terrible, and clothed with horror', he wrote in an account of his subsequent travels, an English

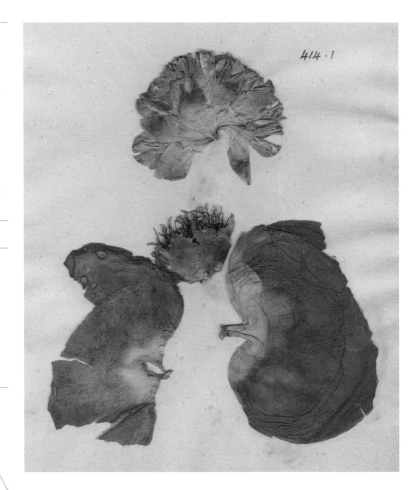

Honourable and far-famed Doctor

I write you this letter being doubtful whether the letter which I wrote to the Cape at the time, will reach you. In my letter to the Cape I explained in detail the reasons which induce me to advise you to come home soon. For God's sake don't think of any further journeys, but rather think of your native country. It necessary that you come home and undertake the office of demonstrator [of botany] at Upsala, as Arch. Linné is on the brink of the grave and mentally and nearly physiologically dead.

... In my letter to the Cape I told as many items of news as I knew. I also said that I found your last letter from Batavia a little obscure and it seemed as if some hypochondria had troubled you. For God's sake reject such hypochondric ideas and try above all to keep your usual cheerfulness, briskness, activity and gentleness.

Peter Bergius, letter to Carl Thunberg, 5 December 1777.

... the plants of the Cape of Good Hope which you have discovered and which form the principal ornament of my collection of living shrubs are part of the trophies which you have erected yourself. Frequently when one of the seeds which reached me from the Cape develops and when consulting its history I find it is you who discovered it, I say to myself 'still another THUNBERG', and I tend the plant still more affectionately.

Nicolaas Burman, letter to Carl Thunberg, 13 August 1823.

I am most anxious to live until you get back; what a joy it would be for me to be present on that great day, and to touch with my hands the laurels that will crown your brow. Lay for me a wreath of flowers on the altar of African Flora.

Carl Linnaeus, letter to Carl Thunberg, 1773.

above right
Masson's original specimen of the plant now known as *Massonia depressa* from the Linnean herbarium.
By permission of the Linnean Society, London

above
Bladder grasshoppers

right
Armoured cricket,
Pachnoda beetles, green protea beetle

edition (to add to the Swedish, Dutch and French translations) of which was published in 1785–86. Sparrman called his narrative *A voyage to the Cape of Good Hope, towards the Antarctic polar circle, and round the world: but chiefly into the country of the Hottentots and Caffres, from the year 1772, to 1776*, a legend that almost dispenses with the need to read the book.

This would be a shame, as he has an enjoyable, jaunty style that conveys keen observation and some sharp social comment. Of course, scientists are serious folk, and this should be the measure of their writing. Sparrman, therefore, was criticised by some of his contemporaries and successors for his light style, perceived lack of accuracy, and long-windedness. Indeed, Sparrman was in the Cape for not much over a year and took 220,000 words to describe his travels of 2,100 km. Thunberg, in contrast, travelled 5,000 km and accounted for them in half as many words as Sparrman. Being even more economical than Thunberg with his quill-pen, Masson described his travels in 12,500 words.

Sparrman was also accused of what we would now call 'sexing up' his account to make it more popular. Maybe, maybe not; but at the end of the day he provides a fascinating and readable insight into life at the Cape in the late eighteenth century.

He was also honourably sympathetic towards the native people and to the slaves. His social comments are perceptive and generally candid, and many of his observations would resonate with us today. He has little time for the pompous (especially if they were overweight; one man he came across 'puffed and blowed more in putting on his stockings than I did when I last went up Table Mountain'), nor the cruel farmers whom he encountered. He relates how one farmer's daughter had 'by some accident or other, already lain in of a black child, the father of which, as a reward for his kindness, had been advanced from the condition of slave to that of prisoner for life on Robben Island.'

An idea of what Sparrman's narrative contains is provided by an introductory list that encapsulated his experiences. This includes, in his words, the following topics:

Floats naked over the river to an islet on a bundle of palmites plants, in order to botanize there.

Adventure of a drunken trumpeter with a hyaena.

Misfortune of a man who harnessed some quaggas [zebras] to his carriage before they were properly tamed.

Almost a whole province intoxicated with a hogshead of brandy.

Divers instances of the sagacity of elephants.

Obliged to live abroad in the air, the walls and every part of the house being covered in flies.

At a rich widow's house is in danger of being kicked out of doors, on his hat being discovered with the brim stuck full of insects.

Locust soup.

Sparrman's travelling companion was Thunberg's erstwhile colleague Daniel Immelman. He suffered from 'very weak lungs' and Sparrman thought the journey would do him good and, put bluntly:

> he had reason to fear a more certain and horrid death in consequence of the complaint he laboured under, than anything that might be apprehended from the attacks of the roving Hottentots or of the wild beasts up the country.

They were, by all accounts, a well-matched pair, and at one stage into their journey:

> formed a resolution not to touch a hair either with razor or scissors, till we should either get into company again with the Christian lasses, or should have an opportunity of dissecting a hippopotamus.

Their preparations for the journey involved acquiring two horses (Immelman's 'a good stallion', Sparrman's 'an ordinary nag'), a wagon and a team of ten draft oxen bought from and driven by 'a boor' (a Dutch farmer), plus the customary accoutrements of the travelling naturalist including brandy for preserving specimens and paper for drying plants. They also took:

> plenty of tea, coffee, chocolate and sugar, partly for our own use, and partly to insinuate ourselves into the good graces of the yeomen, who, by reason of the great distance they are from the Cape, are often without these necessaries.

Not to be forgotten was a good supply of needles 'as by means of these, and a few good words, we should be enabled to gain the good graces of the farmers' daughters, as well as their assistance in collecting insects.'

They set off on 25 July 1775, heading east over the Cape Flats and across the Hottentots' Holland mountain range. Here Sparrman made his first significant botanical discovery in the shape of an unusual member of the protea family. He named it *Paranomus sceptrum-gustavensis*, 'King Gustav's Sceptre', in honour of the Swedish monarch.

right
Pyrgomorphid foam grasshopper
Toad grasshopper

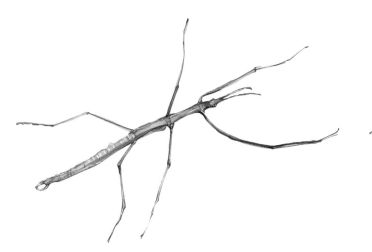

Phasmid stick-insects

They spent some time at Caledon, where Sparrman was able to make observations of the many folk who came seeking a cure for, or relief from, various ailments from the waters of the warm springs there. Rheumatism and gout were common complaints, as well as 'inveterate ulcers … boyls and bondage … epidemical distemper … and indurated tumours'. It seems that immersion more often than not exacerbated their conditions rather than cured them, and he records how by using a salve of wax and honey he healed a girl whose 'leg … was violently swelled and affected with profound ulcers' and for whom long visits to the waters in two successive years had achieved nothing.

After a month, during which time:

the spring together with the month of August made daily advances with her various beautiful bulbous plants, which afterward, when the drought of summer came on, took their leave

they continued east, following the existing route to Swellendam, Riversdale and on to Geelbeksvlei ('Lake of the yellow-billed ducks').

They then headed north over Attaquas Pass and east to Humansdorp, the furthest point reached by Masson and Thunberg. By continuing his expedition, Sparrman went further east than any collector before him and he finally reached the extreme limit of European settlement not far from what is now Somerset East.

Sparrman's narrative so far concentrates on reporting on the people and the animals that his party encountered and the landscape they occupied. Their travelling conditions also add to the atmosphere. Sparrman describes how they often found accommodation with farmers that was not always to their liking because, for example, 'An host of fleas and other inconveniences, to which we were subjected, made us frequently chuse to sleep in the open air …'

He collected many specimens of mammal, bird and insect. Some were preserved as skins while others, notably snakes and lizards, were pickled in brandy, a technique that was not without its problems, as he explains:

I had still reason to fear that the Hottentots would not be able to keep themselves from this delicious liquor, though they knew there was a venomous animal preserved in it … I had long before experienced something of this kind … where a slave had intoxicated himself by drinking some spirits out of a little vessel in which I kept a toad and the foetus of a *hystrix* [porcupine] … Neither could I preserve my brandy from the depredations of my troublesome visitors, till having put several animals

into it, and it being shaken to pieces by the jolting of the wagon, the most inebriating vapours of the brandy, by the assistance of the sun shining upon them, were changed in to effluvia that were highly disgusting, in consequence of the animal particles they contained.

Plants do, however, get a mention every now and then. A notable find on 2 November 1775 was a new species of tree that Sparrman later named *Ekebergia capensis* in honour of Captain Ekeberg 'who was the occasion of my making this voyage; and who by his zeal for natural history, and the great pains he has been at promoting it, is highly deserving of this distinction.'

Having reached the furthest limits of the Cape colony, Sparrman and his companions turned back on 21 January 1776. Following much the same route as they had come, they arrived in Cape Town on 15 April. Three weeks later he had secured passage on a ship and was homeward bound.

A Huge Cape Herbarium

What of Sparrman's Cape legacy? His writing was panned by the scientific critics of the day, and the map that he drew of the region was also given short shrift in some quarters. I would like to think that we can be much more generous in our appraisal. His detractors were essentially competitors and they didn't like his style. I think his account is honest, fascinating, touching, often funny, and full of observation and philosophy that would do credit to someone of twice his age and experience.

In terms of botanising, Sparrman was a keen and efficient collector and, as a doctor, his primary professional interest would have been in plants and their therapeutic uses. As his journeys progressed, however, his interest shifted to fauna rather than flora and in his journal he presents long discourses on the large mammals that he encounters, displaying what we would now consider an unhealthy enthusiasm for shooting many of them.

Sparrman's conversion to mammalogy over botany may, at least in part, have been motivated by the weather. In September 1776 he wrote to the elder Forster that:

The drought was so violent this year, that the like had not been experienced within the memory of man. The most sensible misfortune which the dry season brought along with it was the desolation of the vegetable kingdom. Far from being so fortunate as Dr. Thunberg who has added about a thousand species to the *Flora Capensis*, I found every thing burnt up, and only in the thickest forests met with some perennial plants which were new to me, and which upon a revisal of that gentleman's herbal, I believe are likewise unknown to him.

He then adds, tellingly, 'On the other hand I have been fortunate with animals, and especially in the class of quadrupeds.'

There is some evidence that he amassed more plants than he gives himself, or is given, credit for. Thunberg records how Sparrman 'brought back [to Sweden] a huge Cape herbarium.' Perhaps his specimens, or the majority of them, shared the fate of a large part of his natural history collection that was destroyed by fire. A proportion of the plants that survived may have gone the way of many of his papers which, after his death in 1820, were lost.

On his return to Sweden, Sparrman was elected to the Royal Swedish Academy of Science and was twice its president. He was made a professor and appointed curator of the natural history museum in Stockholm, spending much of his time collating and describing his Cape material.

The final years of Sparrman's life were not happy. Despite his earlier significant contributions to zoology and botany, he bequeathed little to the scientific world over his last 20 years. The reason for this is unknown, but it must have been ill-health or some similar misfortune. In a sad and wholly unbefitting end, he apparently died bankrupt and the auction of his possessions couldn't even cover his debts. He was buried in an unmarked grave.

Anders Sparrman is respectfully and fondly saluted by historians and naturalists, although one, who prefers to cower in the shadow of anonymity, accused him of always seeming to 'make excuses for doing nothing.' His name persists not only in Mount Sparrman in New Zealand, but as the authority of a number of zoological names, including the striped mouse *Rhabdomys pumilio*, and the greater honeyguide *Indicator indicator* (which is a bird, not a driving instruction). His botanical heritage includes *Freesia sparrmannii*, named in his honour by Thunberg, and perpetuations of the poor spelling of the younger Linnaeus, who described the genus *Sparmannia* (Cape Hollyhock) and the species *Erica sparmannii*. But it's the thought that counts, and it is only appropriate that these attractive Cape plants should commemorate this engaging and talented character.

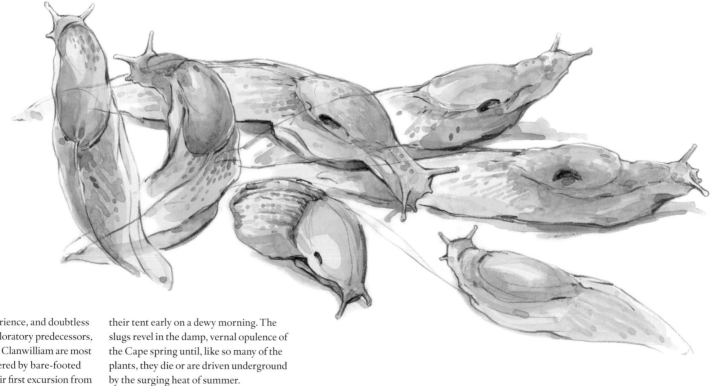

From personal experience, and doubtless that of our more exploratory predecessors, slugs like these from Clanwilliam are most commonly encountered by bare-footed campers making their first excursion from their tent early on a dewy morning. The slugs revel in the damp, vernal opulence of the Cape spring until, like so many of the plants, they die or are driven underground by the surging heat of summer.

Colonel William Paterson (1755–1810) as
Lieutenant Governor of New South Wales,
painted by William Owen (1765–1825) in
c. 1800. Paterson carried out some plant
collecting in Australia, but his major efforts
were at the Cape in 1777–80, his travels
sponsored by the Countess of Strathmore.
Dixon Gallery, State Library of New South Wales

CHAPTER SIX

A View to Gratify a Curiosity

The reputation of the Cape of Good Hope as a place of immense botanical interest was emphatically confirmed by the work of Sparrman, Thunberg and Masson. Two of these figures (Thunberg, in particular) were scientists interested in the taxonomy, distribution and attributes of plants and animals. Masson, while no slouch when it came to his botanical knowledge, was essentially a plant-hunter on the trail of new material to supply Kew and, indirectly, the burgeoning gardening industry in England and the rest of Europe.

Jumping on this flower-laden bandwagon was William Paterson. Although a mere youth of 21 when he arrived at the Cape in 1777, making him the youngest of the Swedish/Scots ensemble that have so far featured (Sparrman was 24, Thunberg 29, and Masson a venerable 31 when they touched-down at Table Bay), Paterson pushed the boundaries of exploration and ventured further into and beyond the Cape than any botanist before him.

Paterson was born in Kinnettles, Fife, in 1755. Not far from Kinnettles stands Glamis Castle, seat of the Earls of Strathmore since 1372 and more well known today as the birthplace of the late Queen Elizabeth the Queen Mother. An earlier member of this aristocratic lineage was Mary Eleanor Bowes, born in 1749 and later to become Paterson's patron. The future Countess was the only child of George Bowes and his second wife, Mary Gilbert. Bowes, a Member of Parliament, had made a fortune from coal-mining interests in the north of England. He was keen to cultivate in young Eleanor the intellect and enthusiasm of his first wife, and the girl did not disappoint. Among the interests that she carried through to adulthood was botany.

George Bowes died when Eleanor was only 11. She and her mother moved to London where, in due course, the now extremely wealthy Eleanor became the focus of attention for various fortune hunters and greedy opportunists. At the age of 16, however, and after various flirtations, she fell for John Lyon, ninth Earl of Strathmore, a handsome but,

by all accounts, lacklustre sort. A protracted period of contractual negotiations followed, and they were married in 1767.

After the social whirl and excitement of her teenage years in London, the new Countess found married life very dull, despite producing five children within six years. During this period, the Earl became consumptive and died while on one of his therapeutic cruises in 1776.

Having bought a smart property in Chelsea, the widowed Countess rekindled her interest in botany. It was at this point that William Paterson entered the picture. At the time he was serving an apprenticeship under William Forsyth, the gardener firstly to the Duke of Northumberland at Syon House, London, then at the Chelsea Physic Garden, and finally St James's and Kensington Palaces. Forsyth is commemorated in the winter-flowering shrub *Forsythia* and was one of the founders of what was to become the Royal Horticultural Society.

It is not clear if the Countess knew Paterson from Glamis, or whether he was recommended to her in London. A further possibility is that Paterson had met Francis Masson at Kew and, enthused by his accounts, had approached the Countess for sponsorship to make a similar plant-collecting expedition to the Cape on her behalf.

Although the coincidence of Scottish geography may have had some part to play in Paterson's appointment as 'her botanical emissary to the Cape of Good Hope', his own version of events suggests that it was either he or someone on his behalf who approached the Countess:

> In this undertaking I account myself particularly fortunate in having been patronised by the Honourable Lady Strathmore, whose zeal for botanical researches induced her readily to accede to the proposal of exploring an unknown country in search of new plants, and to honour me with her protection and support.

Either way, he sailed for the Cape on the East India Company's vessel *Houghton* on 10 February 1777, disembarking just over three months later at Simonstown before the ship continued its voyage to Madras.

'This day made preparation for my first journey which I performed with Robt. Gordon Esqr. second in command at the Cape, and is now Commander and Chief.'

So begins, on 5 October 1777, Paterson's account of the expeditions that he made during almost three years at the Cape. His journal, *A Narrative of Four Journeys into the Country of the Hottentots and Caffraria*, was published in 1789, being the first book (as opposed to an extended journal article, as in the case of Masson; or a translation, as with Sparrman) published in English on the subject. It was accompanied by a series of watercolour illustrations by an unknown artist or artists of plants, people, places, and wildlife.

Paterson's original journal manuscript, the record on which his *Narrative* is based, was lost for many years. Remarkably, however, in 1956 the leather-bound volume was discovered languishing in a box of old sermons in the headquarters of the General Conference of the New Church in Bloomsbury, London. The finder, the Reverend Dennis Duckworth, offered the manuscript for sale to raise church funds. It was bought for £750 by Harry Oppenheimer, the South African industrialist and philanthropist whose collection of Africana also contains Paterson's personal collection of Cape paintings. After almost two centuries, the journal and paintings were reunited. In 1980 the two were published in a single volume incorporating, according to its editors, the 'manifest imperfections due to deficiencies in [Paterson's] education.'

As the journal reveals, Paterson and Gordon's travelling technique was not simply to press on in one party. Often they sent the wagon ahead or by a different route, while the plant-hunters on horseback made diversions or lingered at particular sites where the botanising looked promising. The draught-oxen, meantime, were replaced whenever possible at farmsteads, a fresh team allowing for quicker progress. And right from the start they were not averse to savouring the culinary delights that the country had to offer.

> Day two. This morning we left Cape Town … directing our course southerly along the east side of the Table Mountain towards the place of Mr Beaker where we dined. Here are excellent European fruits. In the afternoon we furthered our journey to a place called Sand Fleet which lies between Constantia and the Bay Falso. Here we were hospitably entertained …

It was not long before wet weather, drifting sands, steep mountains and 'the howling of Hyenas' put paid to what was in danger of becoming an eighteenth-century gourmet tour of the Cape as the travellers headed down the eastern coast of False Bay at the foot of the Hottentots' Holland. Here Paterson found a 'variety of beautiful plants and also the Monsonia in great perfection: Geraniums, Ixias, and Gladiolas and many other plants which were quite new to me'.

Having survived 'several horrible precipices, going by one of which my horse fell with me, but luckily falling in a large Rhus [shrub]', on 13 October they arrived at Cape Hangklip ('Hanging Rock'), a prominent,

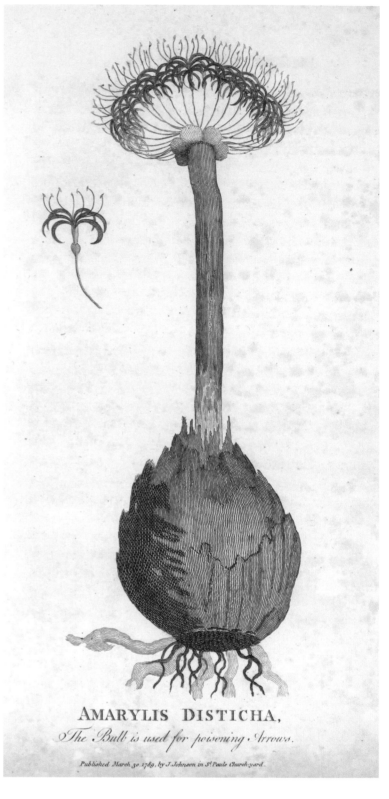

AMARYLIS DISTICHA,
The Bulb is used for poisoning Arrows.

Published March 30 1789 by J Johnson in St Pauls Church yard.

Boophane (formerly *Amaryllis*) *disticha*, a plant which, according to Paterson, 'is called Mad Poison, from the effects produced on the animals which are wounded by the weapons impregnated with it'. If there was any doubt of this, *Boophane* translates as 'ox slaughter'. Paterson had a particular interest in poisons, devoting an appendix to his *Narrative* to an account of various venomous snakes and scorpions and pernicious plants, regretting that 'much that as my chief object was the collection of plants, I had it not in my power to remain long enough in any one place to make such experiments on their several poisons as might have enabled me to have given a clear account of their effects from my own observation.'

right

The type specimen, on which the original scientific description of the plant is based, of *Erica patersonia* in the Kew Herbarium. Paterson discovered his eponymous erica on 13th October 1777 near Cape Hangklip on the east side of False Bay. The marshy flats here remain its main stronghold, although it is under severe threat from coastal development and has already become extinct around Cape Town. A very few plants persist in the Cape of Good Hope Nature Reserve on the southern Peninsula.

far right

'On the tenth [of November 1777] I saw a number of ostriches with which animal this country abounds', observed William Paterson. There are no genuinely wild ostriches in the Cape today, the original population having become extinct through exploitation or hybridisation. Those wandering around the veld nowadays are feral birds descended from crosses with birds imported from further north in Africa for farming.

precipitous peak at the southern tip of the eastern coast of False Bay. Here Paterson discovered 'a beautiful species of Erica with a long spike of yellow tubular flowers.' This was later named *Erica patersonia*.

The weather continued wet and windy so they were doubtless glad of the relief provided by the hot springs near Caledon, a facility that had hosted Sparrman, Thunberg and Masson on their respective travels. Paterson observed that 'There were several people here from the Cape … [with] very different disorders, some with bad ulcers, others with rheumatisms &c, to some it proves effectual, and to others fatal.'

The Barronness of the Country

So far, they had been travelling on horseback, their ox-wagon having gone ahead. The expedition members now met up near Swellendam on 27 October, 240 km east of Cape Town.

As they pressed on eastwards Paterson recorded the landscape, geology and wildlife, particularly plants. He also makes mention of the country folk whom they encountered and who, for the most part, were hospitable hosts as far as their resources would allow, notwithstanding that some of them 'at first … seemed to be shy or rather afraid, who seeing seldom foreigners, they left the house.'

On good days they covered as much as 65 km, on others the going was tougher and they often travelled after dark to escape the heat. When they encountered the rocky ridge mountains and hot valleys of the Little Karoo, Paterson noted the 'barronness of the country.' On 4 November, Gordon records:

> This morning Mr Paterson said he wished to return to the Cape because he was not at all well and feared that he would be unable to withstand the trials of the journey … he took a friendly farewell, which I much regretted as his pleasant personality gave me very much companionship.

Paterson was suffering from acute chest pains and stayed put to recover while Gordon continued the journey.

At this furthest point of his journey, Paterson had reached a place called Beervlei. During his recuperation there and while waiting for a suitable moment, and company, to return to Cape Town, he made some short collecting trips, before making his way slowly back along the route by which he and Gordon had originally come. While still some distance from the Cape, however, his journal grinds to a halt with just the briefest reference to the last two or three weeks of the journey. In his *Narrative* he ascribes this to the fact that:

The wet weather that Paterson encountered during his second expedition would at least have favoured the plants if not the travellers. Soaking winter rains drizzling into spring flood the Cape with flowers, including a multitude of bulbs such as this *Ixia rapunculoides*.

As this country from this to the Cape is well known and described in both Mr. Mason's [*sic*] and Dr Sparrman's narratives, any further account would only be repeating what has been published by these gentleman.

Paterson was back in Cape Town on 13 January 1778. In three months he had covered about 1,500 km and had endured pretty much everything that the country and climate could throw at him. He was doubtless grateful for a rest but, as with most of the Cape plant-hunters, his accounts are largely restricted to his travels and we have little idea how he spent his time while based in Cape Town. This reticence is regrettable, as what to him and his fellow explorers might have seemed mundane domesticity or trivial social detail would have provided us, two centuries later, with interesting insight into life at the Cape. It would also have given us a better understanding of the travellers themselves, although in some cases this is probably what they were keen to avoid.

After a four-month break and with winter well set in, Paterson set off on his second expedition on 22 May 1778.

Pre-eminently Lumpy

Paterson's travelling companion on this occasion was Sebastiaan van Reenen, 18 years old and the 12th child of a leading Cape family.

Their route took them, firstly, east through the Hottentots Holland and on to the coast at the Gourits (a corruption of Goriqua, the local tribe) River estuary. The weather was awful, with snow and hard frost on the mountains and so much rain in the lowlands that they were forced to make lengthy stops to allow river levels to drop so they could cross safely. Some light relief was provided at one spot where 'we amused ourselves shooting wild ducks and water hens which were here in great plenty and so near the house that they were in shot of us without going out of doors.'

They retraced their steps inland until, turning north-west at a place called Plattekloof ('Flat-sided ravine'), their route took them over a series of mountain ranges and intervening valleys and into the Karoo. This is a region that, around 100 years later, the Scottish traveller and author, R. M. Ballantyne, described as '… pre-eminently lumpy', adding that 'Its roads in most places are merely the result of traffic. They, also, are lumpy.' Even in this generally inhospitable and almost inaccessible terrain, isolated homesteads were scattered across the landscape and rarely did Paterson and van Reenen want for accommodation and hospitality at the end of the day.

Despite being very cold, there were flowers aplenty in places. Indeed, winter and early spring, with their associated rains, are the peak time for displays of bulbous plants and annuals at the Cape. A dry winter will see little or no germination, growth and flowering, so wet weather is a price worth paying. Paterson makes note of collecting many beautiful plants and mentions a few by name. His descriptive writing is, however, largely confined to the travails of the journey, the landscape, and fairly frequent reference to the people they encountered, notably the Khoekhoen and their encampments, customs and culture.

By 3 September, the expedition had reached the Copper Mountain, first explored by van der Stel's expedition of 1685. At this point, and for some time previously, Paterson was well beyond the limits of the Cape Floral Kingdom, and he continued further until reaching the Orange River, where there was mutual agreement to return 'having had nothing to eat for two days before except some wild prickly cucumbers which were here in abundance…'

By 8 November they were on the banks of the Olifants River, comfortably back within the realm of the Cape flora. Here Paterson noted a 'variety of plants such as Asphelathes, Licudendrons and many other which were unknown to me.' These are *Aspalathus* (Cape gorse) and *Leucadendron* (conebushes), typical of the sandy plains of the Cape west coast where the party now found themselves.

Ten days later Paterson 'made an excursion to the tops of the mountain where I had a view of the Table Land to the southward, distant about 60 or 70 miles.' The prospect of the Table Mountain massif spurred them on and the party reached Cape Town on 20 November having trekked 2,800 km in almost seven months on the road, or whatever passed for one, lumpy or otherwise.

Paterson undertook two further journeys at the Cape. These followed much the same routes as, respectively, his first journey to the east with Robert Gordon, and the northern one lately completed.

The first of these, his third journey in total, began on 23 December 1778 in the company of Frederick van Reenen, brother of Sebastiaan. That the height of summer was not the best time to go plant-collecting was soon confirmed by the observation that 'passing through this country no variety of plants in flower, everything being quite dried up and hardly grass enough to support the sheep and cattle of the farmers.'

It is no surprise, therefore, that throughout the journey the flora received less attention than observations of people and places. This is not necessarily to the narrative's detriment. Indeed, an uninterrupted list of plants, many unfamiliar to Paterson, would make for a fairly dull read; that the party's slumbers were occasionally disturbed by passing elephants is arguably more interesting. The fact that they saw many other mammals, including lions, African wild dogs, and black rhinos, that have since been exterminated in the Cape or which, in the case of blue buck and quagga, are now completely extinct, adds poignant interest to his chronicle.

Paterson's botanical discoveries do, however, get a mention from time to time, and by this stage the travellers were well into that eastern part of the Cape Floral Kingdom that enjoys summer or year-round rainfall, so there were myriad plants in bloom. On 3 February 1779, he describes the landscape beyond the Great Fish River 'interspersed with a species of Palm [cycad] mentioned by Mr Masson in his second journey.' A few days later:

> we entered a spacious plain, which afoarded us great variety of the most beautiful evergreens I ever saw; and also bulbous plants many of which I found in flower such as Iries [irises], Crynum, one species of which I

found with a crimson flower which for neatness and elegance far exceeded any I had ever seen.

Reaching the Keiskamma River in the eastern Cape by mid-February they turned and made their way back to Cape Town along much the same route. The homeward journey is described in just a few hundred words as it was fairly uneventful and, latterly, 'the country so dry and hardly a plant to be seen.' The expedition arrived back on 23 March 1779, after a trip of some 2,100 km that had taken three months.

'A Variety of Succulant Plants'

Paterson's 'Journey the 4th' took place between 18 June and 22 December 1779, his travelling companion being once more Sebastiaan van Reenen. It being mid-winter, the trip didn't get off to a very auspicious start, being 'detained for three days by the inclemency of the weather', but once they got going, they struck north following the same route and with much the same destination (the Orange River) in mind, as on their previous journey together. In the first three weeks they enjoyed the hospitality of a number of farmers, one of whom 'was glad of our company and begged we would stay several days with him', encountered a succession of rivers in spate that proved hazardous to ford, found mountain tracks washed away by the rains, lost their way, and were delayed by having to repair their damaged wagon.

By the end of July they had reached Ellenboogfontein in Namaqualand, and decided to strike out from there for the Atlantic coast.

> We were much advised by the natives not to proceed, that we had an uninhabited desert to pass where there was neither man nor beast to be seen and great scarcity of water and hardly a blade of grass to support our cattle.

None daunted, they persuaded (with beads and tobacco) one Khoekhoen to accompany them, together with another of the van Reenen clan named Jacobus.

After a few days they arrived at a spot from which they could see the Atlantic in the distance and Paterson 'collected several plants such as Ixias, Gladiolus &c.' They later found 'a variety of plants tho' most of the succulant tribe, so that I could preserve no perfict specimens of them.'

About 15 km from the coast they made camp on the banks of the Buffels (Buffalo) River, remaining for a few days to allow their oxen to rest and 'to range the adjacent fields in search of plants.'

Once on the coast, they found the going strenuous. There was very little freshwater, they failed to catch any fish, and the 'Wild Ducks' that they shot 'were so oily that we could hardly eat them.' This was not surprising, as they were cormorants. They also got lost, and Captain Gordon, with whom they had met up and travelled intermittently, had separated from the party and they 'were in great doubt whether we should ever see or hear of him again.' On 12 August, however, they were brought the good news that a native guide had found a spring and, having been rejoined by Gordon, they set off and found that:

Although Paterson was not the world's most accomplished botanist, he did recognise that the vegetation of the Cape flora, notably its fynbos component, was highly distinctive. One of the defining features of fynbos, the major constituent of the Cape Floral Kingdom, is the presence of restios – reed-like, leafless plants belonging to the Restionaceae. There are more than 300 species in this family, including over 30 species of *Thamnochortus*, such as the one illustrated here. Restios are notoriously difficult to identify, not helped by the fact that male and female flowerheads are borne on separate, often wildly dissimilar, plants.

This place not only afforded us good water but also excellent grass for our cattle, and a variety of succulant plants such as geraniums, Stapalias, Mezm [mesmebryanthemums].

For the next nine days they traversed 'a dry sultry desert where no living animal was to be seen and during which time our cattle had had no water but twice.' On 17 August, reinvigorated no doubt by the discovery of 'an Ostrich nest which contained thirty four fresh eggs; they proved good eating', they arrived at the mouth of the Orange River.

Over the next few days they explored, collected plants and hunted along and across the river. Having stocked up with food, they then retraced their steps to Ellenboogfontein, reaching it 'much animated.' After six weeks traversing 'sultry desarts', Paterson and his companions were delighted to find:

> the whole country covered with flowers of the most beautiful colours, tho' most of them well known such as Ixias, Gladeolias, Geraneums and in the marshy parts varieties of Orchises.

Their route then took them north once again, crossing the Orange River further to the east. A zoological highlight of this part of their journey was the shooting of a 'Cameleopard', a giraffe. The skin was successfully preserved and, as Paterson subsequently noted, 'is now stuffed and in the possession of John Hunter, Esq., Leicester Square.' This is thought to be the first entire giraffe skin to reach England.

Towards the end of November, the party was back into Cape flora country proper, but by this time the spring flowers would have been largely over and the botanists didn't tarry. His fellow travellers having previously gone their separate ways, Paterson arrived back in Cape Town on 21 December having journeyed 2,700 km in six months and five days.

Stoney Broke

While Paterson was busy botanising at the Cape his patron, the Countess of Strathmore, was not exactly enjoying a life of blue skies and sunshine. Following the death of her dull husband she continued an affair that she had begun when he was still alive with George Grey. This didn't last and, despite being pregnant (for the third time, all terminated) by Grey, the following year she married Lt Andrew Stoney Robinson.

This was not a good move. Within a few days of their marriage her new husband, having changed his name to Bowes, discovered that he had no access to Eleanor's considerable fortune, she having drawn up an antenuptial trust barring him from the money. He was not pleased and, by inflicting considerable mental and physical duress, forced her to revoke the deed.

For the next seven years she lived a miserable existence under his control, denied any access to the materials or personages on which she depended to indulge her scientific interests. In early 1785 she escaped and sued for divorce on the grounds of cruelty and adultery. Bowes

caught up with her, however, and had her abducted. She was finally rescued in the north of England, and Bowes and his henchman were given three years in jail and fined £300. The countess was granted a divorce in March 1789 and spent the rest of her rather sad days a victim of wealth and frustrated intellect. She died in 1800.

This upheaval had inevitable consequences for William Paterson. By the time he returned to Cape Town after his fourth expedition, his benefactress was not in a position to send him funds. As a result he was essentially stoney broke (this well-known phrase apparently arising from Stoney Bowes' name) and with debts that could have landed him in prison.

As a genuine casualty of circumstance and, by all accounts, a decent sort, Paterson obviously pressed all the right buttons and, if he did play the sympathy card, then he did it with some virtuosity. He succeeded in borrowing £500 from Colonel von Prehn, the retired garrison commander, and £400 from James Adcock, the valet of William Hickey, an attorney (otherwise described as 'an engaging rake and nabob') returning from Bengal, and with whom Paterson had climbed Table Mountain in July 1777 on Hickey's outward voyage to India. Hickey, in his entertaining memoirs, relates how he met with:

> an ingenious young man, Mr Paterson, a great botanist, who had for several years been employed by that strange and eccentric woman, Lady Strathmore, to go into the interior of Africa for the purpose of collecting rare plants and natural curiosities of every description.

All three were to be Paterson's fellow passengers on the homeward-bound *Held Woltemade* that departed Cape Town on 10 March 1780.

On arrival in Holland, von Prehn demanded repayment. Unable to oblige, Paterson appealed to Hickey and the latter not only stumped up this amount, but Adcock's £400 as well.

A View to Gratify a Curiosity

This was not the last that we hear of Paterson in a Cape context. In December 1870, Britain declared war on the Netherlands and France, and three months later Paterson was to be found aboard an English ship, part of a fleet sent to invade the Cape. Several Dutch ships were captured at Saldhana Bay, but the planned invasion was called off and Paterson continued on his way to India. His mere presence in the enemy fleet, however, raised questions about the reasons for his having been at the Cape for three years. Many of those who had been his close companions and friends during his sojourn felt betrayed. Although there is no real evidence that he had been actively spying, it is likely, and not surprising, that his local knowledge was put to some use when the English fleet arrived off the Cape coast.

Meantime, having served four years in the army in India, Paterson returned to Scotland to nurse a liver complaint. During his convalescence he wrote his account of his Cape travels, including what are, in the context of accusations of spying, the meaningful lines 'I left England with a view to gratify a curiosity, which, if not laudable, was at least innocent.'

In 1789 Paterson married Elizabeth Driver, 'a good cosy Scotch lass and fit for a soldier's wife', and in 1791 sailed for Botany Bay as captain of a company whose job it was to guard the new penal colony there. Over the next seven years he occupied various positions including, for a short time, that of lieutenant-governor of New South Wales, a post he held during a political crisis which saw the arrest of Governor Bligh of *Bounty* fame. He also undertook a number of expeditions, although none on the scale of those at the Cape.

William Paterson died on board ship off Cape Horn on 21 June 1810 while returning to Scotland because of ill-health. If truth be told, he had become an alcoholic, a trait probably nurtured, if not planted, by carousing at the Cape in his early years. A contemporary there, the eccentric explorer and naturalist François Le Vaillant, described how:

> everyone was eager in praise for this traveller, who while he had supplied claret for their entertainment, had shown himself an invincible competitor in the rivalship of smoking as well as drinking.

Paterson had no family ('but had six fine bastards: three in England … this is the case, more or less, with every gentleman in the colony', according to John Grant, an Australian convict), and died poor. He was elected a Fellow of the Royal and the Linnean Societies, but did not fulfil his potential as a scientist having, for example, not written any descriptions of the new plant species that he discovered. As a traveller and collector at the Cape, however, he ranks among the best. This may come as some consolation in view of Thunberg's observations, as recorded on 27 April 1778 in the antepenultimate paragraph of his Cape narrative.

> I met here with a Mr Paterson, an Englishman, who was come to this place, in order to collect from the interior of Africa, and transmit home to his own country, both the seeds and the live roots of such plants, as were scarce and peculiar to these parts. He professed to travel at the expense of some private person, and possessed some knowledge of botany, but was, in fact, a gardener.

Although he must, at times, have been a difficult man to live with, Paterson's wife erected an affectionate and respectful memorial to him in the parish kirk at Kinnettles. This and his plants (not least *Erica patersonia* and *Patersonia,* a handsome genus of Australian iris) serve as more fitting tribute than Thunberg's condescension.

Masson Returns

The late eighteenth and early nineteenth centuries are often referred to, with justification, as the Golden Age of Plant Collecting. British trade and imperialism were at their height, and the collecting and cultivation of exotic plants was visible proof of the wealth and power of the colonial clutches in which Britain held a significant portion of the world.

François Le Vaillant (1753–1824) was exasperated by plant collecting. 'Can it be called preserving a plant to spoil its shape in every part by crushing it flat between two leaves of paper?', he declared. His account of his South African expedition of 1781–84 became one of the most popular travel books of its day, and the discovery in 1963 of 165 paintings commissioned by him provided a wonderful glimpse of the environment experienced by the early explorers.

opposite above
The Khoekhoen trained the horns of their best oxen into decorative shapes.
Library of Parliament, South Africa

opposite below
Dangerous crossing of the Olifants River from Le Vaillant's portfolio. This dramatic incident is described by historian Vernon Forbes: 'Wishing to reach the south bank of the river to hunt elephants, [Le Vaillant] crossed the wide and flooded stream seated astride a tree trunk … His powder-flasks and his watch were slung around his neck, his guns were on his shoulders, whilst, shaded by ostrich-feather plumes, within the crown of his hat was a circular wooden box containing his insect specimens. Tow-lines made of leather thongs attached to the log enabled the Hottentot swimmers to drag it slowly along... Six pages of lively and even at times emotional prose detail the vicissitudes of that protracted crossing from which, by his account, they barely escaped with their lives.'
Library of Parliament, South Africa

The Quagga or striped ass of the interior of the Cape. Once numerous and widespread in the Cape south of the Orange River, Quagga were seen as 'little more than vermin on a large scale' and were shot to extinction in the wild. Chronic overgrazing compounded the situation, William Burchell observing in 1811 that 'sheep had grazed the veld so heavily that scarcely a blade of grass remained' in Quagga country near the southern Cape settlement of Sutherland. Le Vaillant's painting is one of very few depictions of the species. A female that lived at London Zoo from 1851 to 1872 was the only Quagga ever photographed; the last surviving individual died in the Amsterdam Zoo in 1883.
Library of Parliament, South Africa

'Kees', a Chacma Baboon, was Le Vaillant's 'faithful travelling companion'. 'An ape', wrote Le Vaillant, 'is lascivious, gluttinous, thievish, revengeful, and passionate; and if he has not the art of lying, the savages say it is because he does not choose to talk.' Troops of Chacma Baboons still roam the Cape where they have not been displaced by development and agriculture. Their relationship with people is, at best, uneasy and from Le Vaillant's description it is easy to see why – they are too like us. 'If any crime was committed to which gluttony was the incentive, if any theft was discovered, Kees was instantly accused, and the accusation was seldom unfounded'. Kees was also described as 'a good friend, a faithful servant, extremely cunning, fruitful in expedients in times of emergency, and by whom I had more than once been extricated from embarrassments.'
Library of Parliament, South Africa

Crossing the Swartkops River. Forbes notes that 'the peaks of the Groot Winterhoek Mountains in the background … are shown too near or too high, and do not exactly match the individual summits …The artist or artists of these topographical drawings seem to have a predilection for embellishing wild landscapes with near-vertical slabs of stone. Another failing that is evident … is that the trees do not have a characteristic South African appearance, but resemble the neatly trimmed specimens seen in a park in France.'
Library of Parliament, South Africa

Francis Masson had seen some part of this. In 1776, Sir John Pringle 'again petition'd his Majesty … who was graciously pleas'd to consent to Mr. Masson's again undertaking an extensive plan of Operations', and he was off, this time on a four-year collecting trip to the islands of the east Atlantic and the Caribbean.

No journal record exists of the expedition, but some of his letters have survived, including ones to Joseph Banks, William Aiton, and Linnaeus. In 1777 he visited the Azores and then returned to Madeira, thence (in early 1778) to Tenerife. Among the plants he collected and sent back to Kew was *Pericallis cruentus* which, together with subsequent hybrids, is popularly known today as the cineraria, a decorative and popular house plant and, for those lucky enough to live somewhere with a half-decent climate, a border flower.

Whilst on Madeira, Masson wrote to both the elder and the younger Linnaeus. To the father, on 6 August 1776, he apologises for being 'so weak a botanist' because he had made some mistake in the taxonomic designation of a plant. He concludes his letter by:

> imploring the Divine Being to grant you a longer existence on earth, to patronize the great study of the works of Nature; which is the earnest wish of, Sir, your most obliged, humble servant, Francis Masson.

This supplication fell on deaf ears, however, as in his next letter to Linnaeus Jnr, Masson writes 'I condole with you on the death of your great father; a loss which every lover of arts and sciences must ever feel.'

Masson himself, meantime, was 'Still enjoying, though in the afternoon of life [he was 38!], a reasonable share of health and vigour', and in 1779 he set off for the West Indies. Here he offered his services 'both as a Soldier and a Botanist' and soon found himself fighting the French, who had allied themselves with the United States against the British during the American War of Independence. He was taken prisoner in Granada and after his release made his way to St Lucia. Here he was caught up in the 'terrible hurricane' of 14 October 1780, when most of his collection and notes were destroyed.

Resigned to the fact that warfare was not wholly conducive to plant collecting, Masson made his way back to England, probably in November 1781. Two years later he left on a plant-collecting trip to Portugal. Here he took the opportunity to visit Lisbon and its gardens, about one of which he was uncharacteristically scornful: 'the trees and shrubs were of the most common sorts and disposed in extraordinary bad taste.' He again visited Madeira, as well as Spain, Gibraltar and North Africa, before returning to England in 1785.

In October that year, Masson was South Africa-bound once again. Having boarded the British East Indiaman *Earl of Talbot* in Portsmouth, he arrived at the Cape in early January 1786. Here he was to spend almost ten years.

Sadly, no written record (and there surely was one), has survived of this protracted stay. An idea of Masson's activities and movements can be deduced only from his intermittent correspondence, in the form of expenses claims and short letters to Joseph Banks, and by the consignments of plants that he sent to Kew. While these communications, the majority held at Kew and in the Mitchell Library of the University of New South Wales, are of great interest, they provide only a tantalising insight into what Masson was busy with, for busy he was.

Sixty Sorts of Seeds

On arrival, Masson found that the situation at the Cape had changed somewhat during his ten-year absence. Most significantly, his freedom to travel freely had been compromised by the Dutch belief that William Paterson was a spy. It was into this atmosphere of suspicion and mistrust, albeit five years since the brief hostilities, that Masson arrived.

In a communication to Banks of 21 January 1786, he reported that he had taken his letter of credentials from the Dutch Ambassador in London to Governor Van der Graff of the Cape who:

> treated me in the most friendly hospitable manner, but was at a loss how to act respecting my request, as it was ordered by the Company that no stranger hereafter should have liberty to explore the country.

After some deliberation, the Governor acceded to his request to collect plants 'advising me at the same time to conduct myself as not to excite the jealousy of the inhabitants, which was raised to a great degree on account of Mr Paterson.' The bottom line was that, once out of Cape Town itself, Masson was not permitted to travel anywhere that took him within three hours of the coast. This was presumably to prevent him from getting to know the lie of the land and passing strategic intelligence to the British.

As if this wasn't bad enough, Masson was also obliged to inform Banks that 'I shall remember the wine, but am sorry to inform you that it is raised from thirty Rix dollars to eighty . . .' Banks had obviously placed an order for some of the Cape's finest, although he had himself noted when there in 1769 that:

> The famous Constantia so well-known in Europe, is made genuine only at one vineyard … Near that, however, is another vineyard which is likewise called Constantia, where a wine not much inferior to it is made, which is always to be had at a cheaper price.

Masson found himself frustratingly constrained and could not imagine how he was going to maintain, let alone exceed, the standards of his previous Cape expeditions. In fact, he was so fed up that he made a request to Banks to be sent onwards to India where:

> Sir Archibald Campbell [of Inverneil, on his way to take up his appointment as Governor of Madras] and all the gentleman of the *E. Talbot* … gave me assurance of a good reception…had I not succeeded at the Cape.

Agreement to this was not forthcoming and Masson resigned himself to remaining, subject to restrictions that not only included the three-hour-from-the-coast limit but, in addition, only allowed Masson

Seeds of the tall, lanky Bot River Protea
Protea compacta were first collected by
Francis Masson in 1789 in the mountains
near Houwhoek, 100 km or so east of
Cape Town. Within a few years, plants
were brought to flowering at Kew.

scope for searching for and collecting notable herbs and plants on all regions and mountains that lie within the defined limits; provided that also the mountains within the said extreme limits may only be visited and traversed along their lower landward slopes,

essentially so he couldn't get a view of the sea! The eyes of the populace were on Masson and they were instructed to 'immediately arrest him and take him into custody at the nearest place' if he transgressed.

Despite the limitations and uncertainty, less than a fortnight after his arrival Masson sent off a parcel of 'sixty sorts of seeds' to Banks. He reports that he was granted permission by the Governor to visit the Hottentots' Hollands and 'was fortunate to find some of the rarest [ericas] and Proteae in seed.' For a generally laconic character his excitement seems to have got the better of him and he goes on to reveal that he:

also found some new species of Proteae, which is not yet described, and some other Genera, which now convinces me that these mountains, although so near the Cape, have never been properly explored. There is a seed of an *Erica* which I have named *Banksiana*...

Ironically, and to some extent mysteriously, Banks had also decreed that Masson's collecting should be restricted to within a relatively short distance of Cape Town. It was probably a good thing, for Masson, that 18 months had elapsed between informing Banks of his collecting trips and receiving a reply, which included the instruction:

I hope that before this time you have taken up your head-quarters as I directed at False Bay; the most rare plants to be met with in the European herbariums are from that place, & you know that one rare described plant is worth two nondescripts.

Banks seems to be saying 'Anything they have, I want too.'

The real reasons for his decree are, however, open to interpretation on grounds of finance, politics, and espionage, with some authorities considering that botany was, perhaps, the least credible. It's hard to see Masson as a spy, however, as he was too absorbed in his plant collecting. Furthermore, Banks had begun his letter with fulsome praise:

The plants you have sent home have succeeded so much better than any you sent home when you was last at the Cape that we have every reason to praise your industry, & to see the propriety of a search near the place of your residence in preference to expensive journeys up the country, which seldom produce an adequate return in really ripe seeds.

In a postscript to the already unambiguous instruction and, having obviously just received Masson's letters of early 1786, Banks writes:

These letters mention you having undertaken 2 long journeys, which surprised me, as your instructions are very absolute on that subject. What I recommend is a fixed residence during the ripening season at any place where plants are abundant; but more especially that my directions relative to False Bay be complied with; & till you have exhausted that place and [Hout Bay], which I expect will be prov'd rich, I trust you will remain quiet; afterwards you may propose excursions.

above and opposite
Protea seeds and seedheads were common inclusions in the packages that Masson sent to Banks. The group of three are wild almond *Brabejum stellatijolium* fruits, source of van Riebeeck's 'coffee'.

As it transpired, any reprimands were purely academic as the postal system was so prone to delay due to the vagaries of shipping movements, that Masson didn't receive Banks's letter until almost three years after it was written.

Only from Masson's communications with Banks and a few other people, and by the lists of plants received at Kew, can we track subsequent travels. He undertook major expeditions in spring, some taking him beyond the limits of the Cape Floral Kingdom and into the Karoo and Namaqualand. Time was also spent on short trips around Cape Town or longer ones of a few days into the mountains of the south-western Cape.

Masson's expenses sheet for 1788 shows that he made a trip to the Olifants River lasting around two months. The same bill also lists the employment of a Dutch wagonner and a Khoekhoe servant for five months, indicating that he spent a good deal of time 'on the road' even if the trips were shorter. His expenses for 1789 include the costs of a major trip to the Kamiesberg 'being about 400 miles distant from the Cape', and in 1791 his bills show that he undertook journeys to Cape False (Cape Hangklip), Hout Bay, Cape Bona (the Cape of Good Hope proper), and around False Bay.

One way or another, Masson never lost his enthusiasm for collecting, be it on short trips or long. When apologising for not getting consignments sent off he makes the excuse 'the articles are as nearly posted as possible considering my wandering state of living, and you may rest assured that every penney has been employed for the good of the service.' In the same letter of 29 June 1792 he writes:

> I had given up thoughts of making more long Journies but having a waggon I think 4 or 5 months in some of the most unfrequented parts (by Botanists) of the country would be more interesting than remaining about the Cape. In which case I must buy a set of oxen my others was very old and several of them destroyed by the hyaenas. I therefore sold the remaining 8 for 64 Rix dollars to the Butchers.

After months of hauling a wagon over hill and dale at the Cape, these were probably not the most tender of cuts. Tobacco and 'Liquor for the journeys' set him back 26 Rix dollars but, in the absence of many others, we can forgive Masson these minor indulgences, especially as such commodities were also used as gifts for the people he met on his travels.

In August that year Masson journeyed inland to the Klein Roggeveld, an exceptional location for bulbous plants, in particular. Although this route took him some distance east of the Cape Floral Kingdom and into one of the most arid areas in South Africa, its significance lies in the fact that on it he discovered another plant of the genus that was earlier named after him.

Judging by a soft spot for *Stapelia* and for his eponymous genus, Masson appears to be more keen on the obscure, unostentatious species than the showy, elaborate ones, such as proteas, in which the region abounds. With some affection for the man and his flowers, the Government Botanist and President of the South African Philosophical Society, Peter MacOwan, later observed how species of *Massonia* 'attract the dullest eye by their very singularity.'

No opportunity having happened since the date of the above letter I have been enabled to add considerably to the list. I was much afraid that the part of the seeds which was collected last year would not arrive in time to be sown this spring in which case they would have been rather old; but now I am in hope if the ship has a good passage they will arrive in time to be sown. They are contained in two small chests and will be delivered by the India Company Ship Princess Amelia. They contain about six hundred articles.

… A party of Boors has made a journey in to the country of the Cafres in order to search (as they say) for the unfortunate people belonging to the Grosvenor India Ship [wrecked on the Transkei coast on 4th August 1782]. They have however returned without finding any of them: we have not yet heard the particulars of their journey only one of their Captains fell into a pit (which is dug by the Savages to catch wild Animals) and lost his life. Another was trod to atoms by an elephant which he had wounded.

Francis Masson, letter to Sir Joseph Banks, 27 January 1791.

The same morning we went on shore, and were immediately introduced to Mr. Francis Masson. This was a most interesting interview; we had all the botanical news of England to relate, and he had to tell of, and show us, his new acquisitions. This was indeed a treat, more especially in a stroll we took to view his collection in a small enclosed recess under Table Mountain. There we saw, for the first time, the Nymphæa cerulæa, which Mr. M. had just previously added to his collections; many species of Stapelia and Heaths which he had discovered and collected at the same time. He showed us a large assortment of seeds; for a small packet of which we offered a bag of dollars, but he was honourably proof against the temptation.

James Main, 'Reminiscences of a voyage to and from China, in the years 1792–3-4', *The Horticultural Register*, 1836.

Masson on the Bounty

During his time at the Cape, Masson was able to meet many of the travellers and seafarers who were passing through, the majority aboard vessels sailing to or from the Far East. One notable exception was described in a letter of 1788:

> We have in False Bay the Sp. Bounty. Bound to Otaheite [Tahiti] has beat for 30 days off Cape Horn … and at last was obliged to bear away for the Cape of Good Hope. No person belonging to her has yet arrived here but I intend tomorrow to go to visit them.

What a shame that a more detailed record of this does not exist, as it would have been an interesting historical snippet. We do, however, have the observations of Lt Lachlan Macquarie, who arrived at the same time aboard the *Dublin en route* for India, that on one particular evening there was 'a great deal of dancing with the Ladies in the Evening to fine Moonlight on the Quarter Deck.' According to Macquarie's biographer, M. H. Ellis, they were joined by 'Lieutenant Bligh and Mr Mason [*sic*], the famous botanist, who was collecting strange plants for the King in Africa.' Macquarie noted how they all 'staid on board until very late at night'!

What subsequently happened to the unfortunate Bligh needs no repeating here; suffice it to say that the merriment stopped not long after. The next mention of him by Masson is in a letter to Banks of 21 December 1789 in which he reports that he would have sent a consignment of seeds and plants back with Bligh 'but his situation as a passenger on board a Dutch packetboat prevented me.'

With communications and travel between England and the Cape becoming increasingly problematic, Masson considered returning home as early as 1789 and intimated as much to Banks. He was also finding it difficult to send his collections to Kew. The previous year he had written that:

> The English ships has almost forsaken the Cape, & those few that may touch are from Bengal & Madras and crowded with passengers. The China Ships being free of passengers are the best. If I am to return the ensuing year, I should have advice by the first opportunity, and if no English ship occurs that has accommodation, I think a Danish ship would be the next best opportunity for they are very capacious Ships & no passengers, and in general as large as our 74 Gun Ships, providing the Capt. would agree to land me at Dover. These are only hints. I refer to everything for your consideration.

He then goes on to describe how he intends to go on a collecting trip to the Karoo 'among the succulent plants', be back in Cape Town by November, and in the 'following months to pass in the high mountains where the plants come later.'

By the end of 1794 Masson was in a state of readiness should the opportunity arise to get a passage home. He left the Cape for the last time on 17 March 1795.

Hunting seeds was not an occupation confined to plant collectors – they had to beat the locals to the choicest specimens. This is a Cape gerbil, a rodent found only in the south-western Cape. It excavates systems of burrows that are blocked off if a snake or other predator attempts to get in. Cape gerbils can grow to 30 cm long (more than half of that being tail), and feed on seeds, bulbs and roots.

Masson didn't seem to be too upset at leaving, not least because, as he wrote in a letter to Thunberg from his Kensington lodgings on 28 August:

> the whole colony has for some years being falling in decadence and at last almost General State Bankruptsy, having nothing but wretched papper money. It is now fallen into the hands of the English whether they will recover its credite or whether it will remain long in their hands is difficult to say.

Masson also informed Thunberg that he left the Cape:

> with a collection of growing Plants which I have been so fortunate to bring safe home, all my Stapelei (about 30 Spec.) are now growing in Kew garden.

For the next two years Masson was employed at Kew and busied himself with a monograph of his beloved stapelias. Published in four parts between 1796–97 under the title *Stapeliae Novae or a collection of several new species of that genus discovered in the interior parts of Africa*, it contained 41 plates of illustrations of stapelias, almost certainly executed by the author, and was dedicated to the king.

In September 1797, Masson was off on his travels again because, according to Sir James Smith (who bought Linnaeus's library containing letters and specimens sent him by Masson):

> A life of so much leisure soon became irksome to a man who had been used to so much bodily exertion, and mental recreation, amid the wild and novel scenes of nature…

On this occasion, Masson made for Canada where he was as adventurous and apparently no less indefatigable than at the Cape.

The transatlantic crossing saw his vessel boarded by French pirates and the ship's company transferred to another vessel bound for Baltimore, on which they 'suffered many hardships from weather, want of water, and provisions.' They were then taken aboard another ship and arrived in New York after a voyage of four months.

Having explored Niagara and Lake Ontario, Masson then journeyed to Montreal. During the next seven years he continued plant-hunting and amongst the many species he sent back to Kew was *Trillium grandiflorum* which, with its relatives and cultivars, has become a staple of woodland gardens.

Now in his sixties, Masson must have been feeling that time and his constitution were catching up with him. In July 1805, a letter from Banks advised him that he was 'at liberty to return to England on account of his health.' This offer came too late. Francis Masson died in Montreal on 23 December 1805.

News of his death having reached England, it was the younger James Lee who provided an insight into Masson's status among his colleagues. Writing to Sir James Smith, he announced:

Stapelia Gordoni.

Plate 40 from Masson's *Stapeliae Novae*
Masson named this plant *Stapelia gordonii* in honour of Colonel Gordon. It has since been transferred into the genus *Hoodia* but retains its original specific epithet. The plant was discovered by Gordon and an illustration of it appears in his 'Atlas'. Masson did not see the species in the wild and faithfully copied Gordon's illustration down to the last prickle on its prickly stems. This species in particular, and the genus *Hoodia* as a whole, have attained recent prominence through their purported properties as an appetite suppressant. The plant's true effectiveness has yet to be clinically proven, but demands are already high enough to have put the wild populations under pressure.

We are sorry to have to communicate to you the death of our dear friend Masson … we lament his fate most sincerely, he was hard dealt by, in being exposed to the bitter cold of Canada in the decline of life, after twenty five years Services in the Hot Climates, and all for a pittance – he has done much for Botany and Science, and deserves to have some lasting memorial given of his extreme modesty, good temper, generosity, and usefulness. We hope when the opportunity serves you will be his Champion.

In his tribute, Lee also reminds us that:

Masson was of a mild temper, persevering in his pursuits even to a great enthusiasm, of great industry which his Specimens and drawings of Fish, animals, Insects, plants, and Views of the Countries he passed thro evince, and tho he passed a solitary life in distant Countries from Society his love of natural history never forsook him, characters like him seem for the present dwindling in the world, but I trust they will revive…

It was noted that Masson left very little property, and that what he had 'consisted chiefly of the journals of his various travels, drawings, and collections of dried plants and other natural productions.' Smith records how these were left to his two nephews who, in turn, sold them to 'Mr Lee of Hammersmith, a worthy friend of their original possessor.' Lee informed Smith that:

I thought I could have procured the particulars of his life from his Nephews, who inherited his property, but it proves to be *only the journals of his various travels* [my italics] thro the above countries.

The 'above countries' were 'Cape of Good Hope twice, Madeira, the Canaries, Azores, Spain, Gibraltar, Tangier, Minorca, Majorca, the West Indies, and Canada.' Masson's travels *were* his life so it is deeply regrettable that Lee appears to have been placed little value on the journals describing them. I suppose it's too much to hope that they lurk unrecognised in some dusty archive or attic? The discovery of Paterson's journal at least gives faint hope of this.

Masson has been much recorded in biography but because there is so little to go on in terms of primary reference and any chronicle, if there ever was one, of his later travels has avoided discovery, it is difficult to get a new angle on the man. Some myths have been perpetuated in the various accounts, notably that a band of escaped slaves attempted to capture him and use him as a hostage when he was botanising on Table Mountain and he spent the night, either in the open or barricaded in a shepherd's hut, in mortal fear of them. It's true that the mountain was the haunt of escapees (Sparrman, in his words, 'narrowly escaped being plundered by a troop of slaves, that had some time before run away from their masters'), but there is no evidence that Masson also had this experience. One author describes him as a 'naïve and humble gardener … who suffered terrible dangers in introducing 400 flower species alive from the Cape…' Naïve? I don't think so; humble, no doubt. Terrible dangers? In their day, not especially, and nothing that the rest of the population wasn't exposed to, apart from falling off cliffs.

It could truly be said of this Scottish gardening hand [Francis Masson] that he gathered the flowers of the sun in all their radiant glory and in their arid uncouthness. To him fell the spoils of the extreme's of flora's kingdom, from the vividly hued flamboyance of pelargoniums, ixias, gladioli, proteas and ericas, to the bizarre misshapen strangeness of the succulent and cacti kingdom; from the sun-drawn aromas of the scented pelargonium and the belladonna lily, to the evil stench of carrion plants, the stapelias. Green house and border throughout the world owe more to Masson than gardeners ever realize.

Kenneth Lemmon, *The Golden Age of Plant Hunters*, 1968.

There is a commemorative plaque to Masson in the Cruickshank Botanical Gardens of the University of Aberdeen. His most lasting memorials, however, are the genus *Massonia* and various species named after him, his *Stapelia* book, his modest writings, and what have become some of the most popular garden plants in the world. There is also, of course, his venerable cycad at Kew. Visit that and ponder.

Having outflanked the rodents and successfully made collections of seeds, bulbs and living plants, the next challenge was to ship them back to Britain, an undertaking that was far from straight-forward. As this etching illustrates, the 'Cape of Good Hope' may have been the ultimate name of choice for the Peninsula and environs, but Dias's 'Cape of Storms' remained every bit as appropriate. On 5 November 1799 a raging gale led to the sinking of eight vessels in Table Bay, including HMS *Sceptre* with the loss of 349 souls. *Cape Archives*

CHAPTER SEVEN

A Collection of Collectors

The distances covered and the plants amassed by a small but worthy band of pioneer collectors were vital ingredients in the history of the botanical exploration of the Cape. As important, however, was the fact that these 'personalia' (as Professor MacOwan termed them) kept journals for all or part of their sojourns, and some of them wrote scientific treatises on their discoveries. In the case of Thunberg's *Flora Capensis*, this was a milestone in Cape botany.

In addition to these well-documented characters, however, the Cape hosted other collectors whose legacy lies not so much in what they wrote, if indeed they committed anything to paper beyond labels for their specimens and seed packets, but in the plants that they sent back to Europe. Prominent among those who were as proficient with the trowel as they were diffident with the pen were Franz Boos and Georg Scholl.

In May 1786 Boos and Scholl arrived at the Cape, sent by Emperor Josef II of Austria to collect plants for the Imperial Garden at Schönbrunn, his summer retreat near Vienna, where the two were employed.

The gardens at Schönbrunn had, like Kew, developed a more botanical and less recreational emphasis when Josef II's father, Emperor Francis I, dedicated an area of the palace garden to the study of plants. Overshadowed by his wife, the Empress Maria Theresa, he had found himself increasingly impotent in the spheres of politics and finance, and so resigned himself to the role of 'unassuming gentleman' (and father; the couple had eight children, the seventh of whom was Marie Antoinette).

The garden had already been stocked with rare and unusual specimens, doubtless including many Cape species brought from Holland. The horticultural advisers and staff were of the latter provenance and could obtain their plants from the burgeoning Dutch nurseries, 'by which means, in the very first year of its existence, the garden might already be called rich in precious vegetables', according to the newly-appointed head gardener, Richard van der Schot of Delft.

The Botanists of His Imperial Majesty

Although there had always been a circuitous Cape connection with Schönbrunn, it was not until 1780 that this became more direct. In the winter of that year, the elderly head gardener was confined to his bed with gout and one particularly cold night the stoves in the large hothouse died down. To make amends for this oversight, the following night the caretaker who had been left in charge stoked up the fires to such an intensity that the sudden transformation from cold to hot killed most of the plants, or at least those that had not already been frozen. A further blow to Schönbrunn's horticultural potential was the loss of a cargo of tropical plants from Mauritius that perished on the journey. Replacing these casualties of temperature and transport became a priority for Josef II.

Over the next few years, collectors from Schönbrunn journeyed to North and South America and the Bahamas, shipping many rare and exotic species back to the gardens. In 1786, still smarting from the loss

---

opposite

Honeyflowers and Honeysuckers, South Africa

'The Sugar Bush (*Protea mellifera*, Thunb.) is one of a numerous genus of South African shrubs remarkable for their large showy flower-heads ... The slender climber (*Micrlomoa linearis*, R. Br.) entwining it, belongs to the *Asclepiadaceae*.' This classic study by the well-travelled artist Marianne North (1830–1890) depicts malachite sunbirds drinking nectar from sugarbush *Protea repens*, one of many bird-pollinated plants at the Cape.

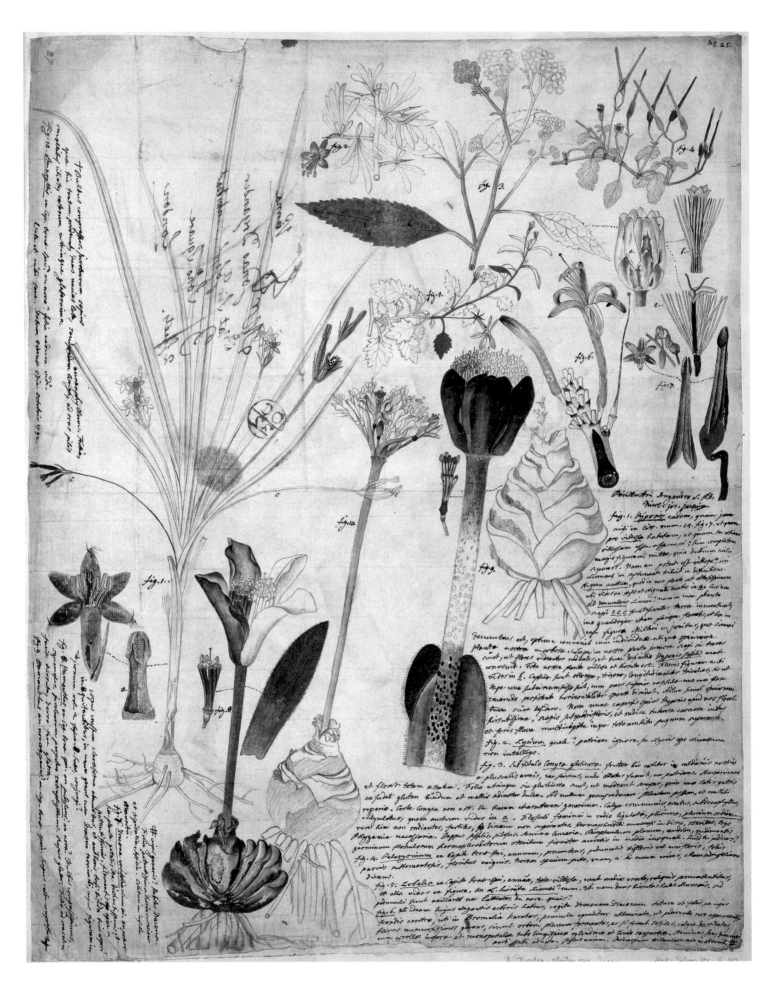

of his plants from Mauritius, and on the recommendation of the garden director, Nikolaus Joseph Jacquin, the Emperor despatched Francis Boos and George Scholl to that island. Their journey took them via the Cape of Good Hope and here they presented their credentials to the governor, being described in the records of the day as 'the botanists of his Imperial Majesty'. After a year, Boos continued the journey to Mauritius while Scholl was to remain at the Cape for the next 14 years.

Sadly, no record of the pair's explorations exists. To get some insight into their travels we rely on passing references by fellow botanists and, more intriguingly, by examining the provenance of the many plants that they sent back to Schönbrunn.

For the first few months Boos and Scholl made short collecting trips in the vicinity of Cape Town, sometimes in the company of Masson and Colonel Gordon (to whom they had a letter of introduction). For the remainder of Boos's sojourn they made some longer trips to the north, probably lasting a few weeks or even a month or two.

In February, 1787 Boos sailed for Mauritius, returning to the Cape the following year with a collection of specimens that extended to 'almost 300' cases. When added to the Cape material that Scholl had amassed there was far too much to send off in one consignment. As it was, Boos was able to take with him on the *Pepiniers* only ten chests of dried plants, seeds, bulbs, stuffed birds, and skins of animals, together with two live zebras, 11 monkeys and about 250 birds.

Boos and his hoard arrived back in Vienna in July 1788. Impressed with what he'd brought, the Emperor slipped Boos 200 ducats and appointed him principal keeper of the Schönbrunn gardens and menagerie, most of which he probably already had in his luggage.

With Boos safely back home, Scholl was left to sort out shipping arrangements for the remainder of their combined collection, to which he added considerably over the next 13 years, maintaining many plants in a small garden that, according to one visitor 'displays not only the taste and integrity of the gardener, but the skill and knowledge of the botanist.' Various interim shipments were made, probably all courtesy of Dutch vessels. A record in the Cape Archives of 8 February 1792 notes that Scholl received 'Permission to send flower bulbs to the Consul of His Majesty the Emperor at Amsterdam.'

Some of these species were new to science and their distributions, often rather small, indicate that Scholl (and Boos during his short stay) travelled extensively in the Cape. The restricted range of one plant, a red-hot poker *Kniphofia praecox* subsp *brucei*, for example, demonstrates that one or both of them travelled as far as King William's Town, almost 1,000 km east of Cape Town. Numerous species of sorrel *Oxalis*, on the other hand, reveal a west-coast route into Namaqualand and onto the edge of the Karoo at Vanrhynsdorp, thence to Garies some 450 km north of Cape Town, and possibly beyond that almost to the Orange River.

Scholl finally made it back to Vienna in 1799, although his prolonged stint at the Cape was not entirely of his own volition. Mention is made of him by Masson in a letter to Banks in March 1789, that:

> The Imperial Botanist has received orders to return, but so little weight has his Imperial Master's name, that he has not been able to get a passage in Dutch ships, and must remain another year, nor has he been yet able to send any of his last year's collection home.

The situation didn't get any better the next year, when Masson reported that:

> The Imperial Collector is still here and has made a fine little collection. But the polite Mynheers [gentlemen] shew so little respect to his great Master, that he is obliged to remain here contrary to his orders and it is with the greatest difficulty that they will receive any part of his collection on board their ships. He remains here under the patronage of Col. Gordon and unfortunately he and the Governor [Cornelis Jacob van de Graaff] are mortal enemies.

On his eventual return to Vienna, Scholl was appointed superintendent of the garden of the Belvedere Palace, a post that he held until his retirement in 1806. It is not known when he died, but Boos died in 1832.

The main legacy of 'The Imperial Collectors' from the Cape point of view is, of course, the plants that were successfully grown at Schönbrunn in the temperate glasshouses built especially to accommodate them. A visitor, Robert Townson, recorded in 1793 how impressed he was with the huge glasshouses and the:

The Cape of Good Hope cannot now be supposed to furnish much of novelty in the department of natural history, especially to transient visitors; but it still continues to afford much amusement and instruction to English botanists. It did so to our gentlemen, who were almost constantly on shore upon the search; and their collections, intended for examination on the next passage, were tolerably ample. They were sufficiently orthodox to walk many miles for the purpose of botanising upon the celebrated Table Mountain; for what disciples of Linnaeus could otherwise conscientiously quit the Cape of Good Hope? In taking so early a departure though it were to proceed to the almost untrodden, and not less ample field of botany, New Holland, I had to engage with the counter wishes of my scientific associated; so much were they delighted to find the richest treasures of the English green house, profusely scattered over the sides and summits of these barren hills

Captain Matthew Flinders (1774–1814), on Robert Brown and Peter Good's disappointment at having to leave the Cape after being able to spend only 18 days there in the spring of 1801 *en route* to Australia. *A Voyage to Terra Australis*, 1814.

immense number of bulbous plants from the Cape, and a rich collection of the genera Arum, Arctotis, Erodium, Geranium, Oxalis, and Pelargonium.

'The gardens', it was also noted, 'were open to the curious at all times, and more particularly on Sunday for the common people.'

£3,200 5s 7d and Three Farthings

It is a pity that so little is known about Boos and Scholl's exploits at the Cape, but they are not alone in being almost anonymous in this department. A similar dearth of information accompanies another important collector, James Niven.

Niven was born in Penicuik, near Edinburgh, in 1776. After working for a year at the botanic garden in that city (for ninepence a day), he took up a post as gardener to the Duke of Northumberland at Syon. In 1798 he was contracted to collect plants at the Cape by George Hibbert, a merchant who had an extensive botanic garden at Clapham. Niven spent from 1798–1803 on this venture before heading back to England.

Three months later he returned to the Cape, remaining there until 1812. On this occasion he was sponsored not only by a syndicate of wealthy and ambitious plantsmen, including James Lee and John Kennedy of the Vineyard Nursery in Hammersmith, but also by Empress Joséphine of France.

It is known that Niven travelled widely in the Cape Floral Kingdom, from Clanwilliam in the north to Albany in the east, and that he sent a wealth of seeds, bulbs and specimens back to England and France. It is also known that on his return to Britain he abandoned everything horticultural and went into business with his brother, John, in Penicuik, making a very respectable living in the financial sector. When he died in 1827 he had assets to the tune of £3,200 5s 7d and three farthings.

Niven's 15 years at the Cape exceed those of all other visiting collectors, but no other details of his travels and works have come to light, which makes it all the more heartening that he is at least commemorated in *Nivenia*, a genus of blue-flowered woody irids. Although Niven collected seeds of *Nivenia corymbosa* (the species from which the genus was described) in 'fissures in rocks by rivulets' near Tulbagh, an earlier but overlooked specimen in the Kew herbarium has an accompanying label bearing the name of Francis Masson.

Nivenia corymbosa was cultivated in major European gardens in the first half of the nineteenth century before vanishing from sight, although it did persist in the Glasgow Botanic Garden until the 1890s. It and the other eight members of the genus are attractive, deceptively tough plants and long overdue a return to more general cultivation.

Although no other plant collectors were to spend as much time at the Cape as Niven or his distinguished predecessors, a flurry of visitors in the first few years of the nineteenth century made relatively fleeting stops of a few days, weeks or a couple of months whilst *en route* to other parts of the world. The majority of these were serious-minded botanists for whom the collection of plants was a scientific pursuit rather than a commercially horticultural one. There were, however, a number of exceptions to the *in transit*, purely academic brigade, most notably William Burchell and James Bowie.

Oct. 17th. Went ashore in morning in company with Mr. Brown, Bauer & Allen and collected a great variety of fine plants, some insects and minerals. The plants for variety and beauty were beyond description. Some I had never seen before, particularly Orchis, drosera & Hemimeris, and many I had never seen in flower, returned on board about 6 p m loaded.

Oct. 18th. Went ashore in morning with Mr. Brown, Allen, Bell and were joined by Mr. Ryeley Surgeon of Lancaster flag ship, a friendly intelligent man and had a long walk over mountains and sandy deserts and collected a great variety of fine specimens of plants, some insects and birds. Mr. Ryeley led us to a house where we found an assembly of Hottentots dancing Scotch reels to Scotch tunes on the violin – returned on board about 9 PM.

Peter Good, diary of his voyage on HMS *Investigator*, 1801.

An accomplished naturalist, artist and writer, William John Burchell (1781–1863) was apprenticed at Kew. Reluctant to join his father's nursery business at Fulham, he travelled to St Helena, thence to South Africa in 1810 where he travelled extensively for five years, amassing a huge collection of natural-history specimens.

The rejected and dejected Burchell arrived at Cape Town on 26 November. For the first few months he lodged with Christian Hesse, a Lutheran minister who hosted many visiting naturalists. He pottered around Cape Town and Table Mountain, but made longer excursions to Caledon and Tulbagh in the company of Hesse and Peter Poleman, a pharmacist. On the last day of one of these trips they were beset by the sort of foul weather that the Cape winter can throw up. The party struggled to find a route across the Cape Flats, being hindered by sand, mud, and overflowing rivers, and floundered back to Cape Town in sodden darkness, Burchell recording that 'the atmosphere seemed bereft of every particle of light.'

Despite, or maybe because of the challenges, these trips inspired Burchell, and his experiences of the landscape and its wildlife set him thinking about the position of Cape plants in the wider context. He noted, for example, how the grain of some *Protea* wood resembled that of:

> cabinet woods brought from New South Wales, which in fact are trees belonging to the same natural order: thus giving to botanists an additional hint that characters may be possibly be discovered in the structure of the wood of plants that may throw some wished for light on a natural classification of vegetables.

Having had a taste of the natural riches of the Cape, Burchell turned his attention towards a bigger scheme and commissioned a purpose-built wagon in which he would undertake a significantly longer journey. His wagon was only about 80 cm wide, the axle width being half a metre more. The basic vehicle cost him £80 but he spent a further considerable sum on fitments, collecting equipment, and supplies.

In the nineteenth century equivalent of a Morris Minor Traveller, Burchell set off from Cape Town on nineteenth June 1811. He seems to have made fairly speedy progress across the valley plains and over the mountains until he reached Karoopoort where, leaving the last outpost of fynbos, he headed out across the vast expanse of the Karoo.

Travels in the Interior

It was to be almost two years before Burchell re-entered the Cape Floral Kingdom, and a further year before he finally arrived back in Cape Town in April 1815 after nearly four years of travelling. During

Bereft of Every Particle of Light

Burchell had a good horticultural pedigree, his father being the proprietor of the Fulham Nursery in London. A career among the plant pots and seed trays did not, however, appeal and in the summer of 1805 he sailed for the island of St Helena. During five years on this remote mid-Atlantic outpost he initially traded in a business partnership, was then employed as a school teacher and, finally, became the official botanist of the East India Company which had a station on the island. Attractive as this job sounds, he was apparently unhappy with his condition of service (and St Helena perhaps does not have a great deal with which to sustain the inquiring botanical mind), and he left for the Cape in 1810.

An additional and perhaps more significant reason for being less than pleased with his lot, was that his fiancée, while sailing out to join him in St Helena, became enamoured of the ship's captain and married him instead.

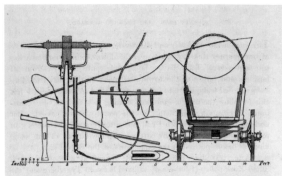

Burchell's *Travels in the Interior of Southern Africa* has been described as 'The most valuable and accurate work on South Africa published up to the first quarter of the nineteenth century.' His talent as an artist, as demonstrated by these illustrations, resulted in many volumes of the early editions of his *Travels* being broken up for their plates.

this time he had penetrated north almost as far as the border with present-day Botswana and east to the Great Fish River where he about-turned on 22 September 1813.

The details of this extraordinary journey are related in Burchell's two-volume *Travels in the Interior of Southern Africa*, published in 1822–24. A third volume would seem to be have been planned, as the narrative stops abruptly at the northern limit of his travels, but never materialised. His subsequent movements can, however, be ascertained from his 14 volumes of field notebooks and plant catalogues, now housed at Kew. These record 8,740 'gatherings' – incidences of plant collecting, the first of which (Numbers 1 to 105) were made between Cape Town and Table Mountain on 5 December 1810. Among the very last were numbers 8,733 to 8,737 'from Mr Hesse's garden in Cape Town', and number 8,740 was a *Satyrium* orchid whose exact provenance is, atypically, unrecorded. He may even have added it to his collection as he made his way, with 48 boxes of specimens, to Table Bay harbour to join his ship for the voyage back to England in August 1815.

Burchell's gatherings combined to produce an astonishing 50,000 pressed plants. A further 10,000 items comprised mammal and bird skins, skeletons, insects and fish, plus more plant material in the form of bulbs and seeds. This is likely to be the largest collection of natural history items ever to have been amassed by a single person in Africa.

Size isn't everything, but this trove represents a remarkable achievement and is made even more scientifically valuable by the annotations that Burchell supplied for the specimens, providing ecological and other details of the habitat in which they were collected.

After 10 years in England, Burchell travelled extensively in Brazil from 1825–1830, again making a large collection of plant and animal specimens, including 20,000 insects. He spent his remaining years in

> My first care this morning was to preserve the botanical specimens which I collected yesterday; and, as I had not the means of pressing them and drying then in the usual manner between paper, I tied them carefully up in a large bundle binding them round as tightly as possible with twine, and wrapping the whole with strong paper. This I left to be sent after me to Cape Town by the first opportunity, intending afterwards to press and dry them properly.
>
> This bundle, however, did not find its way to Cape town till more than a twelvemonth afterwards, and remained in the same state for eight years, when, on unpacking it, every specimen was found to be in as good condition as if it had been dried in the regular manner.
>
> William J. Burchell, *Travels in the Interior of Southern Africa*, 1822.

London. Jilted in his youth, he never married and, latterly, became lonely, introspective and depressed to the extent that he took his own life in 1863.

Burchell is commemorated in a plant genus that contains but a single member – *Burchellia bubalina*, the so-called wild pomegranate of southern and eastern South African forests. He also lives on in Burchell's zebra *Equus burchellii,* and Burchell's coucal *Centropus burchellii*, a close relative of the cuckoos. For a serious naturalist and a compulsive collector on a grand scale, there is no more fitting way to be remembered than in the names of some of your discoveries.

Big Snake and Wildebeest Adventures

Another industrious but ultimately flawed Cape plant collector of the time was James Bowie. The son of a west-end seedsman in London, he became a protégé of Banks and in 1814 he and a fellow-gardener at Kew, Allan Cunningham, were dispatched to Brazil where they spent two years collecting. Cunningham then made his way to New South Wales (which shared top spot with the Cape of Good Hope on Banks's list), while Bowie headed for South Africa, arriving in November 1816.

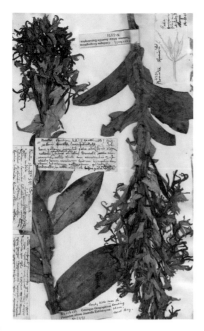

For the first year and a bit, Bowie collected plants in the immediate vicinity of Cape Town and it was not until March 1818 that he extended his range. For the next ten months he travelled extensively in the southern and eastern Cape. Arriving back in Cape Town on 14 January 1819, he devoted the next four months to sorting his collection and arranging for its shipment back to Kew.

He set out on a second journey in early April 1819, again travelling east, this time as far as Plettenberg Bay, and was in the field for another ten months before heading back to Cape Town. On the return journey he was accompanied by George Rex, with whom he had spent three months as a guest on the latter's farm at Melkhoutkraal ('Milkwood Pen', as in livestock enclosure) near Knysna.

George Rex was an engaging character. Born in Whitechapel, London, in 1765, he was the subject of persistent speculation (which he may himself have fuelled) that he was the son of Prince George, later to become George III, and his mistress Hannah Lightfoot. In 1765, however, George was already on the throne and his affair with Ms Lightfoot had, we are assured, long ended. But it made for a good story, and this romantic myth was doubtless more appealing than the fact that he was, more prosaically, the son of John Rex, a distiller, and his wife Sarah.

In 1786, Rex qualified as a notary and in 1797 was appointed marshal of the Vice-Admiralty Court in Cape Town. Once in the Cape, he bought various properties including, in 1803, Melkhoutkraal. Here he became something of a plant collector himself, albeit on a modest scale, and material was indirectly received from him by Kew.

Melkhoutkraal became a traditional stopping-off point for travelling naturalists and collectors until Rex's death in 1839. Certainly Bowie made good use of it and spent some time there during a third major collecting trip in 1820. At the end of another ten months of fieldwork, rather than retracing his steps along almost 700 km of rocky roads, he sailed

above
During the nineteenth century (and no doubt ever since) evangelists heavily outnumbered botanists at the Cape. Tasked by the London Missionary Society to inspect their 11 stations in South Africa, Rev. John Campbell (1766–1840), 'a man of restless temperament and enterprising spirit', made two visits in 1812–14 and 1819–21, later publishing an account of his travels. His crude but charming illustrations here depict a *Watsonia* and 'Mr. Barker's dog'.
National Library of South Africa, Cape Town

back to Cape Town from Algoa Bay. This route was already well established, being a major source of ivory and timber for Cape Town where the supplies of elephants and trees (such few as ever were there) had long since been exploited to extinction.

The hot summer months were based in Cape Town but, by May 1821, Bowie was eastward-bound again, arriving in Algoa Bay at the beginning of June. He then spent six months exploring the northern interior and the south coast before returning for three months *chez* Rex, thence back to Cape Town where he arrived on 4 December 1822.

The following year Parliament reduced Kew's plant-collecting budget and Bowie became the victim of the sort of soft-targeted fiscal prudence that seems never to have gone out of political fashion. He was recalled and sailed for England on the *Earl of Egremont*, arriving in London on 15 August 1823.

Here he was unable to hold down a steady job, spending much of his time, according to MacOwan:

> among the free and easy company of bar-parlours recounting apocryphal stories of his Brazilian and Cape travels, largely illustrated with big snake and wildebeest adventures.

Bowie endured only four years in the drizzly north before heading back to the Cape where he intended to set up a business selling bulbs and natural history artefacts such as stuffed birds. He could probably have made something of this, as he knew the geography and natural history of the Cape as well as anyone and must have had a good network of plant-hungry clients back in England. But he lacked the discipline and the business acumen to be successful and his character flaws proved insurmountable. In Cape Town he 'did no good, being of intemperate habits and becoming involved with bad companions' according to one source, and was handicapped by 'his temper and want of perseverance and tact' according to another. He also didn't like the competition that had sprung up in response to the demands of souvenir-hunting travellers and collectors, writing:

> There is not a snob, a tinker, or tailor, or any other ignorant ass here but is dealing in cats, dogs, and monkeys, and by this opposition to each other, and re-selling of specimens, the prices are raised far beyond their value, considering risk of sea voyage. There is even an officer of the army who has sometimes forty soldiers told off at a time to collect for him.

Bowie spent the next 42 years living an oddly bitter and futile existence, doing little other than complaining about 'ill-treatments, lack of patronage and appreciation.' He ended his days, nominally employed as a gardener under the charitable wing of Henry Arderne, a lawyer and businessman who owned a garden, famous in its day, in the Cape Town suburb of Claremont.

This was a sad and frustrating waste of a man who could have contributed so much to horticulture but clearly went off the rails at some early stage following his return to England (or perhaps beforehand).

Protea effusa was discovered in the mountains near Wellington in 1827 by Johann Franz Drège (1794–1881). He and his brother, Carl, were professional collectors at the Cape, arriving in 1826 and undertaking expeditions that covered more of South Africa than anyone previously. In 1834 Franz returned to Germany with 200,000 specimens of 8,000 plant species. As the first to map the subcontinent's different vegetation types, he can also claim to be 'the father of South African phytogeography'.

He is commemorated in a number of species, including the pretty *Oxalis bowieii*, and in *Bowiea*, a small genus of bulbous succulents, the best known of which is *Bowiea volubilis*, the climbing onion. Not the most flamboyant of plants it can, however, be found in the more esoteric of pot-plant collections.

More famous by far are Bowie's introductions of *Streptocarpus* and *Clivia*. Apart from a recently discovered species, *C. mirabilis* (the 'miraculous clivia' from Nieuwoudtville, some 800 km from its nearest relatives in the Eastern Cape), the latter occurs outside the Cape Floral Kingdom, although it is widely cultivated in pots and flowerbeds both there and elsewhere in the world. The genus was named by John Lindley at Kew for the first person to bring it to flowering in England, Lady Charlotte Florentina Clive, Duchess of Northumberland.

Streptocarpus is sometimes known as Cape primrose, but is not related to the familiar *Primula vulgaris* of European woods and hedgerows. There are species called *S. primuliflora* and *S. primulifolius*, with yellow flowers and primula-like leaves, respectively, but these weren't named until the twentieth century. The common name of Cape primrose, while perhaps having a modicum of justification due to the appearance of the leaf, may equally have been a marketing technique adopted by nurserymen to promote their product using an attractive name familiar to the purchasing public.

This tack certainly succeeded as, after years of hybridisation and

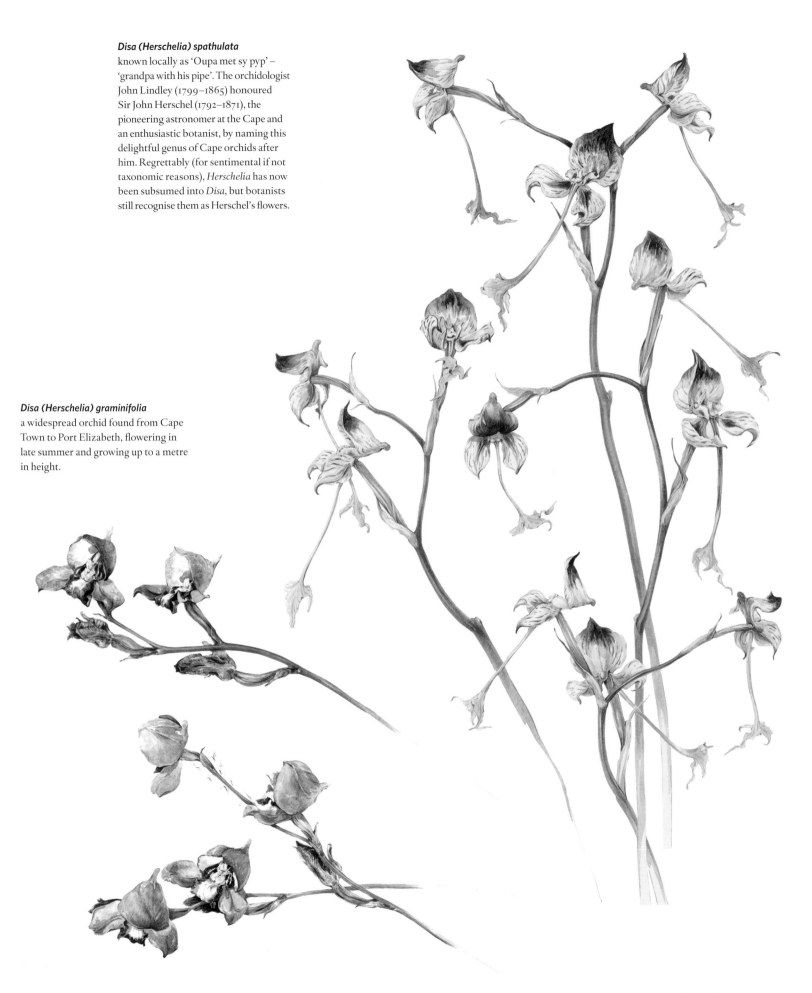

Disa (Herschelia) spathulata
known locally as 'Oupa met sy pyp' –
'grandpa with his pipe'. The orchidologist
John Lindley (1799–1865) honoured
Sir John Herschel (1792–1871), the
pioneering astronomer at the Cape and
an enthusiastic botanist, by naming this
delightful genus of Cape orchids after
him. Regrettably (for sentimental if not
taxonomic reasons), *Herschelia* has now
been subsumed into *Disa*, but botanists
still recognise them as Herschel's flowers.

Disa (Herschelia) graminifolia
a widespread orchid found from Cape
Town to Port Elizabeth, flowering in
late summer and growing up to a metre
in height.

Table Mountain from Salt River (detail)
by Solomon Caesar Malan (1812–1894)
During a short visit in 1839, the Swiss-born, Oxford-educated Malan painted or drew about 90 landscapes and other studies of the Cape. These provide a unique and beautiful record of the environment at the time and allow comparisons with modern Cape Town and its environs. The sandy foreground, now completely urbanised, would once have been rich in bulbs and spring-flowering annuals.
University of Stellenbosch

development, *Streptocarpus* today ranks as one of the most widely grown houseplants. Only two species occurs in the Cape Floral Kingdom but one, first collected by Bowie at Melkhoutkraal and named *Streptocarpus rexii* after George Rex, was the starting point of what is now a major horticultural industry.

It is as well there is something by which to remember poor old Bowie, as he fell far short of the assurance by Banks, conferred on him and Cunningham shortly before their departure for Brazil in 1814, that:

> your reward is within your reach. You will be able to return to your native country before the afternoon of life has closed with a fair prospect of enjoying the evening in ease, comfort, and respectability.

Bowie's 'evening' turned out to be 40 years of poverty and embitterment. He died in 1869.

The Man of God and Geophytes

Someone of stronger character, and who conveniently combined the zeal for plant collecting with one for saving souls, was the evangelist and nurseryman James Backhouse. Always interested in natural history, especially plants, in 1816 he and his brother, Thomas, took over Telford's nursery (established in 1655) in York in the north of England. Their aim was to compete with the London nurseries from which the gentry, even those in the north, were inclined to obtain most of their horticultural supplies. Interestingly, the Backhouse brothers advertised 'Choice Bulbs of Bella-Donna and Guernsey Lilies' in the *Yorkshire Gazette* of 11 September 1819. These were the quintessential Cape geophytes (plants that have underground storage organs, such as bulbs and tubers) of the day.

The nursery flourished, but Backhouse increasingly found that his calling lay in more spiritual ground. This state of mind was probably exacerbated by the death of his wife in 1827 after only five years of marriage. Leaving his two young children and the business in the hands of his family, in 1831 he set off with George Washington Walker, a fellow Quaker, to Australia 'for the purpose of discharging a religious duty.' The pair spent six years there, then three months on Mauritius before arriving at Cape Town on 27 June 1838 to continue their missionary work.

Between 27 September of that year and 11 May 1840 the two covered some 10,000 km on a journey by ox-wagon that took them all round South Africa, visiting mission stations and preaching to anyone who would listen. There was by now a well-established network of missions, religion often being at the forefront of colonisation as much as following closely in its wake.

Backhouse's account of his travels in South Africa runs to 607 pages (plus appendices, and not including his time in Mauritius). If the proselytising was extracted it would be about half this. Nevertheless, it is a

A lantern-slide of James Backhouse (1794–1869) taken during missionary travels in Australia prior to visiting the Cape. *University of Tasmania*

remarkable description by a remarkable man who was 'characterised by earnestness and simplicity, with a tolerant and friendly disposition.' according to his entry in Mary Gunn and Lewis Codd's indispensable *Botanical Exploration of South Africa*.

From the natural history point of view, there are many interesting references to plants and other wildlife in Backhouse's *A Narrative of a Visit to the Mauritius and South Africa*, and he was no mean botanist. The first and final legs of his long journey took him, respectively, east along the Cape south coast, and south down the west coast, so there are some useful observations on the Cape Floral Kingdom. All the while he managed to successfully combine God and geophytes by preaching the gospel and making a substantial collection of bulbs and other material for his nursery as he travelled.

When he returned to England, Backhouse received large quantities of Cape bulbs from Joseph Upjohn, a nurseryman in Rondebosch at the back of Table Mountain. This particular supplier also sent some 4,000 bulbs of 200 species to Joseph Hooker at Kew in 1865, so he recognised a market when he saw one. It is likely that such material was obtained directly from the countryside rather than propagated, an approach that was unsustainable and would certainly be frowned upon, not to say illegal, today. All the same, it represents a relatively early example of the commercial exploitation of plants *in situ* at the Cape by a resident rather than a visitor.

James Backhouse died in 1869, but his family-run nursery at York remained in business until 1955.

Flora Capensis

Within a week of arriving at Cape Town in 1838, Backhouse had 'dined with William Henry Harvey, the Colonial Treasurer, and walked with him through the Kloof, between Table Mountain and the Lion Hill.' He describes how:

> The scenery is very grand. The tops of the rugged mountains to the north and east were covered with snow; but notwithstanding that it was the depth of winter, many beautiful plants and shrubs were in flower.

This was the best introduction to the Cape flora that Backhouse could have asked for, not just in terms of location and plants, but in his choice of companion.

William Harvey was the youngest of 11 children of a prosperous Quaker family from Limerick, Ireland. A youthful interest in natural history was encouraged by his parents and teachers, and while on holiday at Killarney on the west coast he collected a moss, *Hookeria laetevirens*, that had previously only been recorded in the West Indies. Mosses and marine algae, despite their lack (to some) of charisma, were to become consuming interests for his lifetime.

Considering himself 'Neither fit to be a doctor nor a lawyer, lacking courage for one and face for the other and application for both', on leaving school he joined the family merchant business, but spent as much time as he could pursuing his botanical interests. His father died in 1834 and Harvey took the opportunity to reassess his career path.

At this point the post of Colonial Treasurer at the Cape became vacant. The Member of Parliament for Limerick at the time was influential in finding the right person for this job and apparently nominated William's brother, Joseph, for the post. Rumour has it that he actually meant William, but following a change of government and the fact that the name 'Joseph' was inadvertently inscribed in the documentation, the procedure couldn't be amended.

William Harvey (1811–1866) in an engraving from a portrait by Sir William Frederick Burton (1819–1900). Harvey spent only four years at the Cape, but his position of Colonial Treasurer-General allowed him, or he made for himself, plenty of time to explore the botanical riches of the Cape, in every respect a far more worthwhile occupation and one that produced a lasting legacy in the shape of his *Flora Capensis* and *Thesaurus Capensis: or, Illustrations of the South African Flora*.

Harveya capensis. Nine species of the parasitic *Harveya* occur in the Cape Floral Kingdom. They are known locally as 'ink-blom' ('ink-flower') because their petals turn an inky blue-black if pressed or the specimen dries out. The genus was named by William Hooker in honour of Harvey, who responded to the accolade: ''Tis a very lovely plant, with which I am highly pleased and flattered. 'Tis apropos to give me a genus of Parasites, as I am one of those weak characters that draw their pleasure from others, and their support and sustenance too, seeing I quickly pine if I have not someone to torment.'
©*Natural History Museum, London*

Either way, William and Joseph sailed together for Cape Town, arriving there on 17 September 1835. William wasted no time in exploring his new environment and its plants and quickly built up an extensive herbarium.

When Joseph's health began to deteriorate, however, his duties were assumed by William, already his *de facto* deputy. Within a few months they were bound for home again, but Joseph died on the voyage. The post of Colonial Treasurer then fell officially to William and he was back at his desk in Cape Town in late 1836.

Little is known of how Harvey acquitted himself in terms of his official duties, but there is no doubt that he devoted as much time as he could to botany. Most days he would be up at first light to go collecting or beachcombing before the day-job dragged him back indoors. The evenings were spent studying and cataloguing his finds.

With the significant input of other prominent and more well-travelled botanists (Harvey's duties confined him to the city), within a couple of years he had accumulated enough knowledge and material to compile *The Genera of South African Plants*. Published in 1838, Harvey saw this as an introduction to the Cape flora and a stepping-stone to a greater and more detailed work enumerating every species, the first time this had been attempted since Thunberg's masterly *Flora Capensis* of a hundred years before.

Leaving the flat, we crossed a low undulating country thinly clothed with a slight, green vegetation. This is not the flowering season, but even at the present time, there were some very pretty oxalis's & mesembryanthemums, & on the sandy spots, fine tufts of heaths. Even at this short distance from the coast, there were several very pretty little birds. If a person could not find amusement in observing the animals & plants, there was very little else during the whole day to interest him: only here & there we passed a solitary white farm house. … During these days I became acquainted with several very pleasant people:

With Dr A. Smith, who has lately returned from his most interesting expedition to beyond the Tropic, I took some long geological rambles. – I dined out several days, – with Mr Maclear (the astronomer), with Colonel Bell, and with Sir J. Herschel; this last was the most memorable event which, for a long period, I have had the good fortune to enjoy.

Charles Darwin, diary of the voyage of the *Beagle*, June, 1836

Ostrich farming at Groot Post, South Africa
Painting by Marianne North. 'Ostriches are stripped of their feathers twice a year, the operation, it is asserted, causing the bird little pain. Certainly no permanent injury ensues for fresh crops of feathers are produced year after year for a period the limit of which is presently unknown. The trees are Australian gum and wattle, with a hedge of aloe.'
Official Guide to the North Gallery, 1914

left

An exceptional woman by any standards, Marianne North forsook the relative comfort and security in East Sussex for a life of globe-trotting and painting flowers. She visited Cape Town in the spring of 1882 and painted almost 50 studies of the Cape flora (including 'Old Dutch Vase'). She was particularly struck by 'the extraordinary novelty and variety of the different species … the proteas were the great wonder, and quite startled me at first.' Her entire collection of 832 works was donated to Kew and a special gallery was built in the gardens, at her expense, to house them. The accompanying descriptions were published as *The Official Guide to the North Gallery*.

above

Old Dutch Vase and South African Flowers

'This painting done at Groot Post gives some idea of the astonishing wealth [of] bulbous plants of South Africa … At the top, the blue [*Moraea tripetala*], and two flowers of the yellow [*M. bellendenii*]; the crimson flowers above on the right are [a Gladiolus species], and next to it is *Gladiolus* [*orchidiflorus*] … the large star-like rosy flower with a dark centre is [*Spiloxene capensis*]. A yellow *Babiana* hangs over on the left, and below is the pale yellow *Grielum* [*grandiflorum*]…On the right the pale yellow *Sparaxis* [*bulbifera*] and the densely crowded *Ornithogalum thyrsoides*.'

To gather as much material as possible for this project, Harvey had his *Genera* distributed widely and 'sent to resident doctors, clergymen etc, scattered about the country to excite their idle minds to send specimens to Cape Town.' With the patronage of Sir George Grey, the governor of the Cape Colony, and a team of contacts and collectors, Harvey soon accumulated a substantial and valuable herbarium.

Harvey was soon back in Dublin, however, his health having collapsed. He had already taken sick-leave at the end of 1838 but, having returned to Cape Town in 1840, he was only able to continue in post for a year or so before returning north.

His poor health, the magnitude of the task, and new employment in the department of botany at the University of Dublin combined to postpone work on what was to be his *magnum opus* until 1856. The first three volumes of *Flora Capensis* eventually appeared in 1859–60, '62, and '65, co-authored by Wilhelm Sonder, a German botanist and holder of a large collection of Cape specimens.

It has to be said that the *Flora* covered a good deal more than *Capensis* in the phytogeographic sense or the Cape geographically; indeed its subtitle was '*being a systematic description of the plants of the Cape colony, Caffraria and Port Natal*', an area that covered pretty much all of South Africa bar a bit in the north-east.

The scale of the work is demonstrated by the fact that the next instalment did not appear for a further 30 years. Sadly, Harvey had been long dead by this time. Having been laid low in the winter of 1865 with tuberculosis and, despite a move with his wife, a Miss Phelps of Limerick whom he married in 1861, to the relatively amenable climes of Devon, he died on 15 May 1866 while only in his mid-fifties. The reins of *Flora Capensis* were then taken up by William Thistleton-Dyer, Director of Kew, who edited and oversaw the publication of the final four volumes, the last one appearing in 1933.

Harvey's contribution, and the stimulus he gave to South African botany cannot be overstated, despite his relatively short stay in the country. 'All I have taste for is natural history', he once wrote to a brother, Jacob, in North America, 'and that might possibly lead in days to come to a genus called *Harveya*, and the letters F.L.S. after my name.' He achieved both: *Harveya* is an intriguing and attractive genus of parasitic flower with nine species in the Cape flora; and he was elected a Fellow of the Linnaean Society in 1857.

A Minor Deity or Someone's Granny

The nineteenth century, particularly its later years, continued to see more science and less horticulture as far as Cape plants were concerned, stimulated by the work of Harvey and others. To give them their due, the gardening fraternity had, of course, contributed hugely to the growing awareness that there was something very special on the undercarriage of Africa, and their efforts allowed the more academic botanists to recognise and enumerate the extraordinary qualities of the Cape flora.

A prerequisite for understanding something is to give it a name. So

the early collectors, if they weren't all experts in this department themselves, certainly kept the taxonomists busy back home with a constant supply of anonymous plants for scrutiny. Most of the new plants were given functionally descriptive names (*uniflora, prostrata, elongatus, fragrans*, and so on) in the required Latin or Latinised Greek, for which a classical grounding was required. Many others were christened in honour of their original collector, a revered colleague, an aristocratic or royal patron, a minor deity, or someone's granny. *Capensis*, or suitably declined variations thereof, was, understandably, a favourite epithet. I still find this particular one very useful on the increasingly frequent occasions when I can't remember the specific name of some organism from that part of the world. Often enough, I'm right.

Scientific understanding and appreciation of the Cape flora gained momentum in the last quarter of the nineteenth century and positively exploded later in the twentieth. Before losing sight of the work of the early collectors, however, it is appropriate to reflect on just what had happened to the myriad plants, bulbs and seeds that Hermann, Masson, Burchell, Bowie *et al* had unceremoniously expatriated.

A genus of 14 species of dwarf amaryllids, *Hessea* was named in honour of Christian Hesse (1772–1832). A keen naturalist, Hesse hosted many visiting botanists, including Burchell. This is the seedhead of *H. cinnamomea*, a species once common in the seasonal wetlands around Cape Town but now confined to a scattering of sites in the Cape of Good Hope Nature Reserve.

Education is the rage at Stellenbosch, and they had insisted on placing the station a mile off, for fear of disturbing the students! They learnt psalm-singing, if nothing else, and one heard it droning on in every direction (beginning soon after daylight). It was most doleful, and the key dropped full four tones before its last verse was over. I could not help thinking of the Rev. H.J., who warned me before starting that 'Africa was a most untidy country, and I should find missionaries littered all over the place.' I breathed more easily when I escaped from that Calvinistic settlement, and the flowers seemed to get brighter and more abundant as I approached Table Mountain.

Marianne North, *Recollections of a Happy Life*, 1892.

CHAPTER EIGHT

From Cape to Cultivation

When Bartolomeu Dias and his weary crew made brief landfall at the Cape of Good Hope in 1488 there wasn't much reason for them to collect plants. If they weren't good to eat (and rather few Cape plants are), then taking up precious space on tiny ships with perishable souvenirs was hardly justified. There was certainly no question of attempting to transport live specimens as there was barely enough water on board to supply the crew, let alone a bunch of weeds. Even the 'thistle from Madagascar' that made its way to Europe almost 120 years later was likely to have been a consequence of curiosity rather than strategy.

As logistics and fashion finally converged at a point at which plants were collected as commodities in their own right, be it for garden or herbarium, their provenance mirrors that of human exploration, discovery and colonialism. In the late sixteenth and early seventeenth centuries most of the foreign species arriving in Britain and western Europe had come from Near Eastern countries such as modern-day Turkey, where French merchants traded at the Mediterranean ports. For the remainder of the seventeenth century Canada and Virginia were the primary sources. In the late eighteenth century and thereafter, Australia and wider North America became important. From the 1820s until the end of that century North America continued to provide new material, along with Japan, and in the early twentieth century Western China opened up to provide a bountiful supply of material for the garden.

The first botanical garden was established in Pisa in 1543, followed by that at Padua, Italy, in 1545. An English equivalent was founded at Oxford in 1621 and by 1648, under the directorship of Jacob Bobart, it contained some 1,600 plant varieties. Flowers were by now also being grown in pots ('portable gardens') perched on windowsills and elsewhere inside and outside the house, a fate that awaited a multitude of Cape plants.

The Golden Age

It was not until 20 years after the establishment of Jan van Riebeeck's settlement that plant collecting at the Cape progressed beyond the opportunistic and into something resembling systematic with the visit of Paul Hermann in 1672. Nevertheless, even a modest level of collecting had allowed Cape plants to appear in the botanical literature as early as 1605. By this time the Dutch were already at the forefront of European horticultural development, a position made possible by the wealth that flowed from the expansion of their trading empire and, not least, by inquiring minds and internal politics.

In the late sixteenth century the Low Countries had for some time been under Spanish Rule. In 1568, the seven northern provinces of the northern Netherlands signed the Union of Utrecht, bringing together the Protestant north against the Catholic south and its Spanish overlords, and precipitating the Eighty Years War. In 1581 the united provinces declared their independence from Spain and, fearful of losing vital trading ports, notably Antwerp, Philip II of Spain bolstered his military presence in the area. Antwerp was reclaimed by the Spanish in 1585, by which time many of its wealthy Calvinist merchants had fled north, converging on Amsterdam. This influx of prosperity and intellect made Amsterdam a seat of commerce, industry, and learning, and botany and horticulture were prominent among the pursuits indulged in by the prosperous, educated and leisured class.

The founding of the VOC in 1602 can be attributed to the ambitions and imagination of the immigrants and, in no small way, to the invention by Cornelis Corneliszoon of a mechanised sawmill; this enabled windmills to power up-and-down blades that could speedily reduce trees to

opposite

Cupid Inspiring Plants with Love
One of Francis Masson's most celebrated introductions, *Strelitzia reginae* quickly captivated English gardeners, at least those who could afford one. Being such as unusual plant, it was soon commonly invoked in classicism and fantasy. Here the god of love and erstwhile troublemaker takes a pot shot at an unsuspecting *Strelitzia* in *Temple of Flora, or, Garden of the botanist, poet, painter, and philosopher* by Robert Thornton (?1768–1837), an elephantine volume printed on elephant paper (58 × 71 cm) published in 1812. Two other plates feature Cape plants – an 'Artichoke Protea' (*Protea cynaroides*), and a 'Group of Stapelias'.

planks for ship-building on an industrial scale. Within a few years the VOC fleet exceeded 100 vessels,

Amsterdam prospered, and the Dutch entered a 'Golden Age' that lasted the better part of a hundred years.

Seeds, Roots, Cuttings and Shrubs

Plants were a significant feature of this prosperity. Most significantly, the impact of Clusius and the publication, in 1601, of his *Rariorum Plantarum Historia* fuelled the horticultural fever. He it was who introduced and developed many of the bulbs, including the tulip, from which the Dutch horticultural industry has reaped rewards ever since.

The VOC now ruled the commercial waves, establishing highly lucrative trading routes to the East Indies and elsewhere in Asia, with spices high up the shopping list. The company also had exclusive access to Japan, a concession denied other would-be traders. Vessels heading to or from the East rested their crews and took on freshwater, meat and firewood at the Cape. As long as they monopolised this route, the Dutch were the most frequent callers at the Cape and, in consequence, the prime collectors of the region's plants for introduction to Europe.

Collecting plants had, in fact, become part of the seafarers' mission. In 1720, the pre-eminent physician of the time, Hermann Boerhaave, recorded that:

> Few captains, whether of merchant or warship, left our ports without specific instructions to collect seeds, roots, cuttings and shrubs wherever possible, and bring these back to Holland.

The gardens, both public and private, of Leiden and Amsterdam (where botanic gardens had been founded in 1587 and 1682, respectively)

were the prime beneficiaries of this practice, and European botany as a whole benefited to unprecedented levels from what amounted to a passion for pepper.

While the Dutch were busy accumulating impressive collections of Cape and other exotic plants, an upsurge of interest in such things occurred in England in 1688 when James II fled the country to exile in France, leaving the throne to be occupied by his daughter, Mary. She was married to William of Orange and together they jointly ruled Britain for five years. Their main residence was Hampton Court Palace, to which Mary brought her rare plant collection from the couple's famous baroque garden at Het Loo near Apeldoorn in the Netherlands. Here, tender species were grown in large containers or pots of metal or Delft pottery, and in the summer taken out of their hothouses and into the garden.

The cross-Channel connection established, a steady stream of Cape species arrived not only in Holland but, consequently, in England as well. Many of the new plants fell straight into the hands of the king and queen and their loyal accomplice Willem Bentinck. Bentinck, later made Earl of Portland in recognition of his services, also developed his own garden at Bulstrode in Buckinghamshire, part of a portfolio of handsome properties bestowed on him by a grateful monarch.

Mary died (of smallpox) in 1694, and William in 1702, but their Dutch influence and, to some extent, the Cape species they introduced, lived on. During their reign, gardening in England, according to Stephen Switzer, nurseryman to the nobility, 'arrived at its highest perfection.'

Stoves of Eighteenth Century England

By now, not only were ornamental plants achieving a high public profile, but the whole idea of their cultivation and breeding was gripping

Ever since the time of its first settlement the Cape has been a constant source of pleasure and delight to the botanist and the gardener. Though Cape plants have somewhat gone out of fashion of late years, it is probably still true that no single country in the world has contributed so largely to European conservatories and gardens as the Cape of Good Hope … From 1775 to 1835, Cape plants may be said to have been quite the rage. The conservatories, temperate houses, and gardens of England and the continent teemed with the Pelargoniums, Heaths, Proteas and other handsome flowering shrubs, and the lovely bulbous plants of Irideae, Amaryllidae and Liliaceæ; and the pages of the Botanical Magazine and other similar periodicals were filled with figures and descriptions of them.

Harry Bolus, 'Sketch of the Flora of South Africa', in *Cape of Good Hope Official Handbook* 1886.

The first botanical garden in England was established at Oxford with Jacob Bobart (1599–1680) as its director. He was succeeded by his son, also Jacob (1641–1719). In 1753 Linnaeus named the genus *Bobartia* (rush irises) in their honour. It is most noticeable in its natural setting when its long, tough leaves conspire to trip unwary walkers. The bee-pollinated flowers last only a day or so and, being unpalatable to stock, a field full of these is a sure sign of over-grazing.

sectors of society. In response to this the implements and materials required to cultivate them, and books on how best to do this, appeared on the market by the barrowload.

In 1724 a volume that set the trend and the standard for horticultural and gardening books for the next 150 years and more was published. This was Philip Miller's *The Gardener's and Florist's Dictionary*, a practical guide for novice and *cognoscenti* alike. It was based on Miller's own experience as a market-gardener's son, as a commercial florist at Southwark and, most importantly, a gardener at the Chelsea Physic Garden from 1722–70. He introduced the first Cape lobelia that, as a bedding or trailing cultivar, was to become such a feature along with pelargoniums in early nineteenth century gardens, whether big, small, or municipal. Miller's *Dictionary* ran to a number of editions, that of 1752 (the sixth) being the starting point for the naming of garden cultivars.

Miller's *Dictionary* also describes something that was essential for the cultivation of the great majority of Cape species - the 'stove', being 'contrivances for preserving such tender exotick Plants, which will not live in these northern Countries without artificial warmth in winter.' As early as 1635 a visitor to the Chelsea Physic Garden had observed that:

> What was very ingenious, was the subterranean heat conveyed by means of a stove under the conservatory, all vaulted with brick, so as he has the door and windows open in the hardest frosts, secluding only the snow.

A stove was a greenhouse heated by a fireplace that sent hot air through serpentine flues built into a back wall. The moderate and dry heat combined with good ventilation was a pretty good, if serendipitous, emulation of the natural conditions under which plants grew at the Cape. So this set-up suited them perfectly and with the arrival of more material from that part of the world, Cape plants, for a time, dominated the stoves of eighteenth- and nineteenth -century England.

The early stoves did suffer from a lack of light, although not much could be done if the sun didn't shine as there was, of course, no artificial illumination at the time apart from candles and lamps. These were pretty ineffective and potentially damaging as far as plant welfare was concerned. Some Cape plants, therefore, suffered from yellowing and didn't look their best.

Over the years, improved greenhouse design was able to address this shortcoming by incorporating more glass. This was facilitated by the invention of cast-iron which, by the beginning of the nineteenth century, was becoming a major building material. This allowed for narrower supports and, therefore, more space for glass without compromising the strength of the structure. With the invention of the cylinder process of manufacture, glass became available in much bigger sheets, particularly after 1832 in England.

From Cape to Kew

While a gentle flow of Cape plants made its way to England, either directly from source or via Holland, it was not until the last quarter of the eighteenth century that the taps were turned to release a torrent. This was, of course, achieved through the efforts of dedicated collectors, of whom Masson, under the auspices of Joseph Banks at Kew, was the first.

It must have been a hugely exciting time at Kew to receive unassuming brown paper packets. In 1793 a recording system was instituted that logged the arrival of material, and the first entry in the new records book, on 7 June, documents '82 boxes and tubs' received from Botany Bay. The second entry was made the following day and reads:

> Rec'd 3 Boxes of Plants and one of Seeds by Sir Joseph Banks Bart. The Plants appearing when opened to be all dead, they came packed in moss

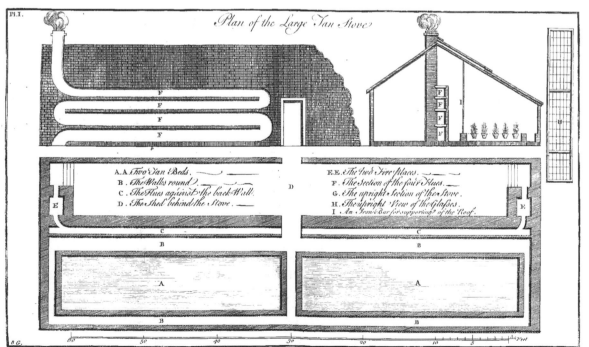

'Plan of the Large Tan Stove' from Philip Miller's *Gardener's and Florist's Dictionary* of 1724. This important development in heating glasshouses allowed Cape plants to be successfully cultivated in England.

BARK. The refuse bark which has been used for tanning leather, and which produces considerable heat by its fermentation … The plants in pots are generally plunged in it, at first to half the depth of the pot, and afterwards to the rim. Substitutes for bark are stable-dung, leaves of trees, chaff, and any other vegetable or animal substances which ferment in decaying.

Mrs Loudon, *The Ladies' Companion to the Flower Garden*, 1849.

1793.

6.

The second entry, dated 7 June 1793, on the first page of the Kew record book details a consignment of seeds and plants sent by Masson from the Cape.

7. June.   Rec.d from Botany Bay 82. Boxes & Tubs.
            brought from thence by his M. Ship Atlantic
            & supposed to be a present to the Garden from
            Gov.r Phillips ――――

18 ..   Rec.d 3 Boxes of Plants & one of Seeds by
            Sir Joseph Banks's Bark the Plants appear
            ing when opened to be all dead, they
            come packed in Moss ― The Box contain
            ing Seed is from Masson & has in it
            several Parcels of Specimens & which
            are to be shewn to Sir J. B. the first
            time he comes to Kew .

18   Rec.d N3 ― from Sir Joseph Banks
            a Box contain.g Seeds & Roots collected
            by Mr. Menzies ― N. Holl.d &c.
            amongst which curious Banksias &c
            ― N.2. a Parcel of Seeds. containing
20.   Papers sent by Capt. Paterson .

Annuals, succulents and shrubs from the Cape were enthusiastically cultivated in Europe in the late eighteenth and early nineteenth centuries. While some went on to become staples of the garden others, despite their good looks or quirky characteristics, fell from grace or favour. Clockwise from top left, *Lobelia valida*, *Salvia africana-lutea*, *Felicia amoena*, *Ruschia sarmentosa*.

At Kew every weed of the veld seemed to be in blossom in the houses – all labelled with long Latin names which gave them an air of distinction and set them apart for our homely 'Vygies' and 'Buchus'.

Dorothea Fairbridge, *Gardens of South Africa*, 1924.

– the Box containing seeds is from Masson and has in it several parcels of the Specimens which are to be shewn to Sir J. B, the first time he comes to Kew.

This was the first of many Masson entries and, thankfully, not all his material was past redemption when it arrived (although the 'dead plants' may have been dormant bulbs). Indeed, so much was being successfully received that a new 18 m glasshouse was constructed in 1792 to accommodate his expanding Cape collection, prominent among which were heaths and succulents such as mesembryanthemums.

While some of Masson's plants went on to become perennial favourites, a high proportion were unlikely to persist beyond a generation or two when it was found that they were not suited to cultivation or were simply not that attractive. As always, enthusiasts with a taste for the esoteric might have been happy to grow some of these out of curiosity, but to become more widely popular among ordinary people, who did not have grand glasshouses and battalions of staff, a plant had to be both good-looking and easy to cultivate.

The Primordial Potplant

Among the species least likely to enjoy a future in the gardens or on the windowsills of the masses was the cycad *Encephalartos altensteinii*, collected by Masson in 1773 and received at Kew in 1775. Still alive and well, it qualifies as one of the world's oldest pot-plants. A prickly customer it has, to the uninitiated, rather few redeeming features apart from being of great historical and botanical interest (fossil specimens

have been found, dating from 280 million years ago). Not the speediest grower, the Kew plant was 30 cm high in 1822, 91 cm in 1848, 2.43 m in 1908, and 3.35 m in 1961, the stem by then somewhat stooped and tottering. Sir James Smith notes that in 1819 Masson's cycad 'produced a male cone, which, being considered remarkable, led Sir Joseph Banks to come and see it, such being his last visit to the garden.'

This venerable pot-plant certainly long outlived the peak of Cape plants at Kew. In the first few years of the nineteenth century they went into a decline, and even Masson's Cape greenhouse was commandeered for the profusion of species arriving from the new colony of Botany Bay. So it was out with the old and in with the new, even if the 'old' had not had a great deal of time to settle in or realise its potential.

The gardens in general were also suffering from a lack of finance and interest, Kew having been deprived of the support of George III following many years of illness and blindness culminating in his death in 1820. In that same year Kew was 'robbed of its most important and influential friend' in the figure of Banks.

The gardens had entered a period of scientific and horticultural lethargy such that, according to William Bean (curator in the 1920s), 'the fortunes of Kew had sunk to their lowest degree' by 1837. It was not until 1841 and the appointment of Sir William Hooker as the new director that it began to reclaim its position as the finest botanical garden in the world. Head-hunted from his position as Professor of Botany at the University of Glasgow, Hooker's contribution to Kew was enormous and under his 24 years of skilful direction the garden, its contents and activities were revitalised.

The Flower Garden Displayed

Although Kew may, for a while, have been well-endowed with Cape plants, the appearance of many of them in print brought the *flora Capensis* to a wider, if somewhat elite, audience. A major disseminator of information was *The Botanical Magazine: or, flower-garden displayed, in which the most ornamental foreign plants, cultivated in the open ground, the greenhouse and the stove, are accurately represented in their natural colours*. The brainchild and product of the horticulturist William Curtis, the first issue appeared in 1787 at one shilling a copy. A decade earlier, Curtis had published *Flora Londinensis*, an account of the capital's wildflowers. Although a fine publication it did not succeed commercially. Familiarity obviously bred contempt, or something approaching it, in the gardening world, as it was new foreign plants that captured the interest and imagination of Curtis's subscribers. His *Botanical Magazine*, in contrast, with its pages of must-have exotics, prospered. Curtis himself edited the magazine almost up to his death in 1799 and, despite experiencing a few hiccups in its 220 years, it is still in print today.

A multitude of Cape plants has appeared in Curtis's magazine. They were particularly numerous in the early years of the publication when those collected by Masson ('this laborious investigator of nature', as Curtis described him) were brought to flower at Kew and by various nurserymen and other enthusiasts who had managed to obtain seeds or plants, often through the good offices of Joseph Banks. The magazine certainly provides a reflection of their abundance and the interest that they generated. For the majority of Cape species it was also the first and, until the advent of modern field-guides and other local publications, the only time that they had appeared in print and were illustrated.

The first 20 volumes of *Curtis's Botanical Magazine* contain 786 plates, of which nearly a third are devoted to Cape species, the majority courtesy of Masson. The first to feature, as Plate 20 in Volume I, was *Geranium (Pelargonium) peltatum*, the illustration displaying a species that, despite more than 200 years of horticultural tweaking, would be easily recognised today by anyone with a 'trailing geranium' in a window box or hanging basket.

above
Of the 36 plates in the first edition of Curtis's magazine, three feature Cape plants, the first being this Ivy-leaved Pelargonium *Pelargonium peltatum*. If Curtis's magazine can be taken as a reliable indicator of the rate of introduction of Cape plants into England, a peak was reached in 1802 when 38 of the 48 species featured in volume 16 were from the Cape. Between 1787 and 1811, the golden age of plant collecting in the region, 450 Cape plants appeared in the pages of the magazine.

left
In figuring *Strelitzia reginae* Curtis confessed to having 'deviated from our usual plan' in giving it such extensive coverage, including a fold-out illustration, in his magazine in 1790. The species was named by Banks in honour of Queen Charlotte, her pedigree being the House of Mecklenberg-Strelitz. Now known more commonly as the crane flower or bird-of-paradise, it is a hugely popular ornamental in warmer parts of the world, including California where it has been declared the city flower of Los Angeles.

Many of the plants featured in Curtis's magazine were obtained from specialist nurseries that had sprung up in response to the burgeoning interest in all things horticultural. Soon to be particularly celebrated for their Cape plants were William Rollison's Springfield Nursery at Upper Tooting, Surrey, especially noted for its Cape heaths, and James Colvill of the King's Road Nursery, Chelsea, founded in 1783 and to become famous for its bulbs.

Perhaps most prominent among the nurserymen of the time, however, was James Lee. Born in Selkirk in the Scottish Borders in 1715, at the age of 17 he left home and walked to London to seek work, his journey being interrupted by catching smallpox at Lichfield.

Lee came from a humble background. His parents were:

> respectable … but not in the station that allowed them to give him any further education than is in the power of everyone to attain in that part of Britain, and which, at that periods, was generally superior to what those of that rank in England can arrive at.

This may tell us something about why England was at this time awash with Scottish gardeners, many of whom rose to the highest levels of their profession. The social conditions in Scotland at the time also encouraged many young people to seek a better life elsewhere. Selkirk was a very poor district and 'subjected to ecclesiastical domination … arrogant and intolerable … there was nothing into which ministers did not push their meddling fingers', according to the local nineteenth-century historian, Thomas Craig-Brown.

Accounts of Lee's early career in London are often contradictory. It would appear, however, that he worked for the Duke of Argyll at his property at Whitton Place near Twickenham (the Duke also allowing Lee to continue his education through free use of his library), for the Duke of Somerset at Syon (which later passed to the Duke of Northumberland), and for Philip Miller at the Apothecaries' Garden at Chelsea before becoming a partner of Lewis Kennedy at the Vineyard Nursery, Hammersmith in about 1745.

Kennedy was already an established nurseryman by the time Lee arrived on the scene, but together they built up the business until only Kew could rival it in the variety and extent of its plants. Being on the main road from London, the nursery was well placed for passing trade and for transporting orders to their destinations in the country. The nursery was also visited by many prominent botanists of the day, including Thunberg, and among its overseas customers was Thomas Jefferson, third president of the United States.

A further draw of the Vineyard Nursery was the fact that Lee had become something of a gardening celebrity. Despite his almost complete absence of a formal education, he made a free translation of the *Philosophia Botanica* of Linnaeus, publishing it in 1760 under the title *An introduction to Botany containing an explanation of the theory of that science and an interpretation of its technical terms extracted from the works of Dr. Linnaeus and calculated to assist such as may be desirous of studying that Author's methods and improvements*. While it would be unfair to term this 'Linnaeus for Dummies', it essentially brought the great man's work to a level that was accessible to the general public.

There is no doubt that Lee was assisted by more learned individuals with this task, as he readily and graciously acknowledges, but it was still an impressive achievement. He also corresponded with Linnaeus and held him in considerable awe as the 'Father of Natural History'. Linnaeus obviously had respect for Lee, too, and named the tropical plant genus *Leea* after him.

Although a commercial concern, the Vineyard Nursery readily exchanged material with other gardens and collectors, not just in London and elsewhere in England, but with Schönbrunn and the botanical gardens of Paris and Strasbourg. Indeed, it was reported that:

> Lee might have died rich, but he was notoriously generous, and cared not what expenses he was at for the attainment of rare plants, and when he possessed such as might have procured him a golden harvest, he chose rather to give duplicates away to lovers of Botany, before selling them to the rich but careless collectors of flowers, rather led to them through ostentation than from a laudable enthusiasm in the pursuit of knowledge.

Lee died in 1795, his share of the nursery passing to his son, also James, who later became the sole owner. The Vineyard Nursery remained in the Lee family until the 1890s.

A strong Cape connection with Lee and Kennedy's nursery is demonstrated by the fact that their nursery catalogue of 1774 contains 54 different mesembryanthemums. Their South African suppliers included William Paterson and John Pringle, the latter an agent for the East India Company who sent *Ixia* and *Gladiolus* bulbs, together with other plants. Among the many Cape species that the nursery introduced into cultivation were *Ixia crateroides*, *Gladiolus floribundus* and *Watsonia brevifolia*.

Lee's partner in the Vineyard Nursery, Lewis Kennedy, was a less demonstrative figure and seems not to have had the same impact upon the horticultural world. He died in 1782, his son John taking over the partnership until he sold his share to Lee in 1818.

John Kennedy was twice married, his first wife, Margaret, having 12 children and his second, Ann, nine. One of the daughters by Margaret married the botanical artist and horticultural entrepreneur Henry Andrews. Kennedy wrote much of the text that accompanied Andrews' *Botanist's Repository*, a publication that was established to rival Curtis's magazine and which appeared monthly between 1797 and 1814. Andrews is best remembered for his illustrated works on Cape heaths and pelargoniums.

Of Kennedy's children by Ann, one was christened Joséphine after the Empress of France. The cynic might interpret this as a move to ingratiate himself in order to bolster her custom – in 1803 alone she bought plants from the nursery to the tune of £2,600. The Empress and Kennedy were, however, more than just shopkeeper and customer, as they collaborated in the syndicate that sent James Niven to collect plants at the Cape.

Amaryllis Josephine. *Amaryllis de Josephine.*

The Garden of the Empress

It is hard to imagine the Empress Joséphine wandering round the stalls and bazaars, but it is said that she picked up a bulb of the *Brunsvigia* species that now bears her name 'while on a shopping expedition to Paris.' Whatever its secondary origins, it flowered in her garden at the Château de Malmaison on the outskirts of Paris and was named in her honour by the botanist and her artist in residence Pierre-Joseph Redouté.

Joséphine became Empress in 1804, and although the childless couple divorced in 1809, she retained her title and lived in Malmaison, a property that she had bought in 1799 to be her home with Napoleon Bonaparte.

It is understandable that Joséphine should seek solace in her garden, as during their relatively short marriage she didn't see much of her husband, his work tending to keep him away from home for much of the time. One obituarist observed that she was 'Extremely unhappy during her husband's reign [and] she sought refuge from his roughness in the study of Botany.' She also had a competitive streak and if there was a particular plant in another garden that she didn't have, she wanted it, and what she wanted she generally got. She was, however, in the time-honoured tradition of most, if not all, gardeners, a generous dispenser of horticultural largesse, which all combined to create a beautiful and rarity-filled garden at Malmaison.

Brunsvigia josephinae from *Les Liliaceae* by Pierre-Joseph Redouté. One of the largest of the Cape geophytes, its bulb can measure 20 cm in diameter, the stalk and umbel of 30 to 40 flowers reaching 65 cm. Nine species of *Brunsvigia* occur at the Cape, some of them taking well over a decade to reach first flowering and living for 100 years or more. The genus was named by Lorenz Heister (1683–1758) in honour of the House of Brunswick, specifically the 'enlightened despot' Karl Wilhelm Ferdinand, Duke of Brunswick-Lunenberg (1735–1806). Heister's reference was a bulb of *B. orientalis* originally christened, with reason, *gigantea*, sent from the Cape by Governor Tulbagh in 1748.

Joséphine obtained Cape material not only through contact with the likes of the Vineyard Nursery, but also as an unlikely consequence of the war between Britain and France. It is reported that Joséphine demanded that any plants among the spoils from captured English ships should be sent to her. Conversely, and curiously, such was the *bonhomie* between society gardeners on either side of the Channel, that the British Admiralty instructed their commanders that should any French ship be captured that had aboard it consignments of plants destined for the Empress, then their delivery to her should be assured. It is also on record that John Kennedy, *'célèbre cultivateur'*, was provided with a passport that allowed him to travel through Europe during the Napoleonic Wars, without let or hindrance, in order to obtain plants for the Empress.

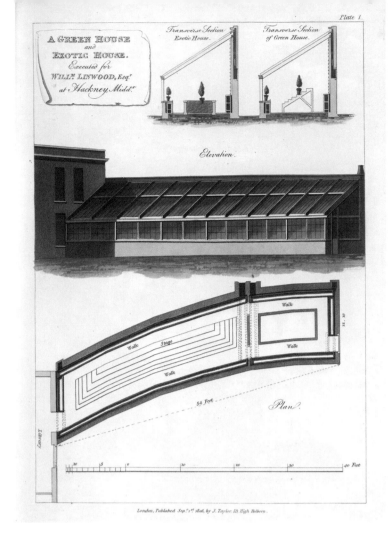

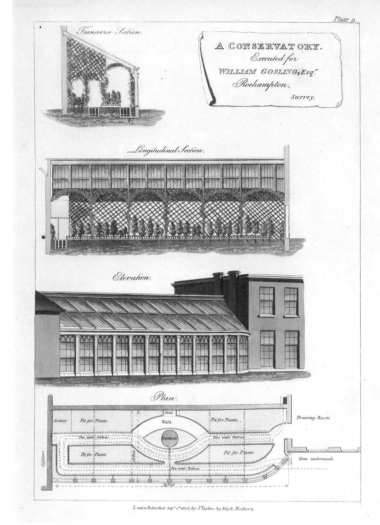

Joséphine spent much of her time increasing and consolidating her plant collection at Malmaison, a passion not universally shared by those around her. One of her ladies-in-waiting, Georgette du Crest, wrote that:

> When the weather was fine, the green-houses were inspected; the same walk was taken every day, on the way to that spot the same subjects were talked over; the conversation generally turned upon botany, upon her Majesty's taste for that interesting science; her wonderful memory which enabled her to name every plant; in short, the same phrases were generally repeated over and over again, and at the same time, circumstances well calculated to render those promenades exceedingly tedious and fatiguing. I no sooner stepped onto that delightful walk, which I had so admired when I first saw it, that I was seized with an immoderate fit of yawning.

Although the garden is best remembered for its roses, many plants not previously cultivated in Europe flourished at Malmaison. The description of the gardens as containing:

> eucalyptus, hibiscus, phlox, camellia, numerous varieties of South African heaths, myrtles, and geraniums, cacti and rhododendrons, certain kinds of Dahlias, not to mention rare tulips and full hyacinths

points to important contributions from the Cape.

The best evidence, however, for the variety of Cape plants that Joséphine was cultivating is to be found in the exquisite paintings of Redouté. He also designed the greenhouses in which many of these plants grew, but it is his portrayals of the latter that have survived. In particular, a magnificent range of Cape bulbs is depicted in *Les Liliacées*, published in 80 parts, the last of which appeared in 1815, the year after Joséphine's death.

Redouté also worked at Kew in the late 1780s, providing the illustrations in the form of stipple engravings for *Sertum Anglicum or Rare Plants which are cultivated in the Gardens around London, especially in the Royal Gardens at Kew*. This was the project of the Parisian magistrate and botanist Charles L'Héritier who, enthused by the diversity and rarity of new plants arriving there, visited the gardens to see them for himself. A number of Masson's introductions featured in this publication.

Coming as it did just before Waterloo and the collapse of the French Empire, Joséphine's death in 1814 from pneumonia (the result of a walk in cold weather round her beloved garden), led to the flowerbeds, greenhouses and water features of Malmaison falling into decay; only a fragment of them remains. *Brunsvigia josephinae*, however, still thrives in the rocky fringes of the Cape Floral Kingdom where it comes into spectacular, leafless bloom at the driest, hottest time of year.

Cape Plants in Print

In the way that certain plants, such as Joséphine's *Brunsvigia* (which was rare and difficult to grow in captivity), were to remain the preserve of the well-off, publications such as *Curtis's Botanical Magazine* and the

opposite
By the early nineteenth century, providing
the facilities for growing exotic plants had
become big business. These designs are
from George Tod's *Plans, elevations and
sections, of hot-houses, green-houses, an
aquarium, conservatories, & c., recently
built in different parts of England, for
various noblemen and gentlemen: including
a hot-house and green-house in Her late
Majesty's gardens at Frogmore,* essentially
a sales catalogue published in 1823. In
their day, these structures would have
been full to the brim with Cape plants.

right
Protea coronata (green) was introduced
to England in 1822 and is arguably more
unusual than it is attractive. Sugarbush
P. repens (pink; a pale yellow form also
occurs) is one of the most widespread
proteas at the Cape. It was introduced
to Kew by Masson and flowered there in
about 1780. It has all but disappeared from
England but has been grown in Californian,
Australian and New Zealand gardens
since the late nineteenth century.

The frontispiece from Elizabeth Kent's
Flora Domestica depicts a well-endowed
greenhouse in which, amongst other
plants, a Cape pelargonium, an erica
heath and an aloe flourish beneath a
fruitful vine. The book is subtitled 'The
Portable Flower Garden with Directions
for the Treatment of Plants in Pots and
Illustrations from the Works of the Poets'
and was published in 1831.
RHS, Lindley Library

others of its genre were not aimed at working-class gardeners of modest
means and small gardens. This niche was to be filled to overflowing
by *The Gardener's Magazine and Register of Rural and Domestic
Improvement*. First published in 1826, its objective, in the words of
J. C. Loudon, its editor and chief contributor (or 'conductor' as was the
style of the day), was:

> to disseminate new and important information on all topics concerned
> with horticulture, and to raise the intellect and the character of those
> engaged in this art.

John Claudius Loudon was born at Cambuslang, Lanarkshire, in
1783. He trained at a nursery near Edinburgh before moving south to
London in 1803. Here he set himself up as a garden designer and
adviser, specialising in the relatively small gardens that were beginning
to spring up in the fashionable London suburbs where gardening on
the grand scale was simply not possible. Such city gardens were, however, able to display features and plants, including a number from the
Cape, that reflected the owner's wealth and status. Most importantly,
in Loudon's mind, was the need 'to supply exercise and recreation for
the occupier and his Family.' Garden therapy is nothing new.

In 1830, Loudon married Jane Webb and together they became a
formidable horticultural and literary team.

John had been debilitated in his youth through rheumatic fever and
his right arm had to be amputated. Jane became not only his amanuensis,
but an independently talented botanical artist and productive writer.
Her list includes such titles as *Young Lady's Book of Botany*, *Ladies'
Companion to Flower Garden*, and *Gardening for Ladies*, the latter selling
1,350 copies on the day of its publication in 1840. The beautifully illustrated *The Ladies' Garden of Ornamental Bulbous Plants*, published in
1841, is full of Cape species, demonstrating just how popular these
plants were at the time. Other Cape bulbs featured were the height of
fashion at the time but have since been lost from general cultivation.

'C.B.S.'

In the same year that Loudon launched *The Gardener's Magazine* (1826),
Robert Sweet published his *Hortus Britannicus*, a 'Catalogue of Plants
cultivated in the Gardens of Great Britain arranged in Natural Orders.'
Sweet was another enterprising botanist, nurseryman and publisher,
and his 'Catalogue' is an exhaustive list of all the species and cultivars
known to be growing in British gardens at the time. An impressive, if
arcane work, it reflects just how much material was coming into the
country from virtually all corners of the world, not least the Cape.

Once you've got your eye in, the best way to pick out the Cape species
in Sweet's list (which runs to 492 pages) is to look for 'C.B.S.' (*Caput
Bonae Spei*), in the 'Native places of growth' column. Hardly a page
goes by without this trio of initials making an appearance and it takes
up many in their entirety.

In fairness, it has to be said that 'C.B.S.' in those days described an
area that extended well beyond the Cape of Good Hope proper and

The popularity of Cape plants resulted in them decorating not only gardens and greenhouses, but textiles as well. This nineteenth century dress incorporates Cape ericas, daisies and mesembryanthemums.
©*Victoria and Albert Museum, London*

covered a greater expanse of the country than that which we now recognise as the Cape Floral Kingdom. C.B.S. more accurately represented what was known as the Cape Colony until 1910, when it became the Cape of Good Hope Province (or simply Cape Province) with the Union of South Africa, and remained as such until regional reorganisation in 1994. Indeed, Sweet's 'C.B.S.' would have been less a reflection of political and botanical boundaries and more a manifestation of the footsteps of those early plant collectors who had headed away from Cape Town and the coast and into decidedly dusty and uncharted territory. Like Harvey's *Flora Capensis*, it would have extended to the northern limits of South Africa.

Nevertheless, for historical, logistical, and botanical reasons, today's Cape Floral Kingdom was the major contributor to the impressive total of 'C.B.S.' plants in *Hortus Britannicus*. The list is, therefore, a fair representation of the remarkable variety of Cape plants that had arrived in Europe by the early nineteenth century.

To be included in Sweet's list the plants must have been growing somewhere in the British Isles at some time. His sources of information were any nurseries and botanical gardens that he could visit, mostly in and around London, and whatever publications were available at the time, such as Curtis's magazine. Many Cape species, by all accounts, might have been represented by some ephemeral pot-plant, soon to be lost without trace not long after it struggled to put on a few leaves or, if it was lucky, a flower, before keeling over and being consigned to the compost heap of horticultural history.

Sweet lists over 3,600 species and almost 350 genera from 'C. B. S.' Featuring prominently are 90 species of pelargonium, together with almost 450 of their cultivars, demonstrating just how much this particular genus had captured the horticultural imagination (and market) in a relatively very short time. Remarkably, there are also more than 460 ericas and 281 mesembryanthemums (of which only one is a hybrid). Of the 36 gladioli, only one ('Colvill's') is a hybrid, giving no indication of how this group would explode exponentially onto the floral scene in a few years time. There is also a healthy sprinkling of species with which today's gardener would be familiar, particularly bulbous types such as *Agapanthus, Ixia, Sparaxis* and *Zantedeschia*.

A significant proportion of Cape plants on Sweet's list are, however, unlikely to be found in any shape or form in modern gardens. You won't, for example, see many greenhouses or flowerbeds today bedecked with Large Pubescent Cotyledons, Tetragonal Adenandras, Long-peduncled Fig-Marigolds or Starfish Stapelias.

Prominent among those that have persisted and flourished, however, are those that found themselves a horticultural patron. One such plant was the *Gladiolus*, and its promoter was an important pioneer of plant hybridisation, William Herbert.

The Spectre of the Tomb

The third son of the Earl of Caernarvon and Lady Elizabeth Wyndham, Herbert's first profession was that of a lawyer. He then served as a

Member of Parliament before becoming a cleric. In 1814 he was appointed to the parish of Spofforth in Yorkshire where he was rector for 36 years before being installed as Dean of Manchester.

When Herbert wasn't writing gothic novels (perhaps the most memorable being *The Wizard Wanderer of Jutland* and *The Spectre of the Tomb*) or carrying out his clerical duties, he was a passionate gardener and plant breeder. He undertook pioneering and incisive research into plant hybridisation, notably of bulbs, and wrote extensively on the subject. In his monograph *Amaryllidaceae*, published in 1837, he reveals himself as an inquisitive and skilled scientist, asking all the right questions and finding the answers, or at least attempting to, through systematic experimentation and the recording of his techniques and results. One of the plant groups that he grew in his Yorkshire garden and on which he focussed particular attention was the Cape gladioli (see chapter 9).

Most Cape plants, meantime, still required the luxury of a greenhouse, a facility that was becoming increasingly sophisticated with advances in technology. The design and construction of glasshouses certainly occupied the mind of John Loudon. He speculated upon the structures that could be built with a new material, cast-iron, writing:

> There is hardly any limit to the extent to which this sort of light roof might be carried; several acres, even a whole country residence, where the extent was moderate, might be covered in this way, by the use of hollow cast-iron columns...

He foresaw glasshouses 'of a hundred and fifty feet from the ground, to admit the tallest oriental trees and the undisturbed flight of appropriate birds among their branches.'

Loudon was ahead of his time, but by 1851 Joseph Paxton had designed and built the ultimate glasshouse – the Crystal Palace. Paxton was, in his way, the gardening celebrity of his generation in the way that the Loudons were of theirs. Born in Bedfordshire in 1801, he started work as a gardener when he was 15. After seven years in the service of various landed gentry, he found work with the Horticultural Society in Chiswick. There, apparently, he impressed the Duke of Devonshire, from whom the Chiswick gardens were leased, and was offered a post at his Chatsworth estate in Derbyshire for 25 shillings a week and a cottage.

He and the Duke became great chums, together making the Chatsworth garden one of the most remarkable in the country. An impressive variety of Cape plants was grown here under vast areas of glass, notably 'The Great Conservatory' that Paxton designed and on which work began in 1836.

Many of the Great Conservatory's Cape contents were provided by Baron von Ludwig, who had his own 'noteworthy botanic garden' in Cape Town. He had visited Chatsworth in 1838 and, the Duke recorded, 'was so charmed with its conception that he stripped his garden at the Cape of the rarest produce of Africa.' Not all of von Ludwig's gifted plants would have been native to the Cape, however, as his garden was stuffed with exotics from the world over. One way or another, however,

ROYAL BOTANIC SOCIETY, REGENT'S PARK.
The first great fête of the Society for the present year took place on the 24th ult ... Orchids were arranged on the lower courses of terraces, as were also Pelargoniums, which were numerous, and with few exceptions excellent ... Cape Heaths, too, scattered over the banks in lavish profusion, helped to contribute to the gaiety of the scene...

The Florist, 1854.

Sir Joseph Paxton (1801–1865) embodied the spirit of the Victorian age in his grand designs and technological innovation. His 'Great Conservatory' at Chatsworth held one of the finest collections of Cape plants. He describes his first day as follows:'I left London by the Comet coach for Chesterfield and arrived at Chatsworth at half-past 4 o'clock in the morning of the 9th of May, 1826. As no person was to be seen at that early hour I got over the greenhouse gates by the old covered way ... I then went down to the kitchen garden, scaled the outside wall, and saw the whole of the place, set the men to work at 6 o'clock ... and afterwards went to breakfast with poor dear Mrs. Gregory [the housekeeper] and her niece; the latter fell in love with me and I with her, and thus completed my first morning's work at Chatsworth before 9 o'clock.' Portrait by Octavius Oakley (1800–1867).
©National Portrait Gallery, London

the Cape's representation at Chatsworth in the 1840s included 98 species of heath and 92 mesembryanthemums.

The Greenhouse Effect

For a while, this sort of cosseted cultivation, from modest urban greenhouses to stately homes with their great glasshouses, was all the rage. The decline of all but a favoured few Cape plants indoors, however, took place quite quickly. Not only were they victims of changes in gardening taste and fashion, but of technology. The dry, well ventilated, stove-heated greenhouses had suited them perfectly, but as heating systems and irrigation improved it became possible to create a warmer and more humid environment under glass.

In 1826, William Whale invented a system that employed water heated by boilers and passed along cast-iron pipes. This certainly suited tropical species, and it was aimed at just that – the great masses of orchids, ferns, bromeliads and so on that were hitting the horticultural shelves in the way that Cape plants had done decades before. But it was the kiss of death for the great majority of sea-and-mountain-breezy, drought-revelling Cape plants and they simply could not survive side by side with lush novelties from steamy jungles in even steamier glasshouses.

Sir Joseph Hooker bemoaned this development and lamented the new generation of gardeners 'trained in the pernicious art of overwatering in hot-houses which is suited mainly to the culture of tropical American plants.' Like any conservative grumpy old man, he didn't like change, but he also was genuinely sorry to see the demise in cultivation of so many beautiful Cape species, the majority of which had not had their horticultural mettle fully investigated or developed.

A Poor, Ginger-bread Entity

Holding their own, however, were certain Cape plants that had lent themselves to production and hybridisation on an industrial scale and, being cheap and cheerful, had caught the imagination of the gardening and non-gardening public alike. Nevertheless, human nature, being perverse as it is, dictates that any plant (or person, for that matter) achieving any level of popular success should become a target for disparagement. One such plant was the *Pelargonium*, and one such critic was Shirley Hibberd.

Hibberd was pre-eminent in what might be termed suburban horticulture and, like the Loudons before him, catered for a clientele that included the middle classes with their small gardens and indoor plants. His first horticultural publication was called *Town Gardening*. Appearing in 1855, it was the first of many of this *genre* and added to a list that included *Rustic Adornments for Homes of Taste* and *Clever Dogs and Horses*. In 1884 he became the first editor of *Amateur Gardening*, a title that continues to this day.

In *The Amateur's Flower Garden* of 1871 he describes the techniques and tastes of the age, grumbling that:

During the last twenty years there has been a constantly-increasing tendency to superficial glare and glitter in garden embellishment. The magnificent displays of bedding plants in our public parks and gardens have, without question, favoured a false estimate of the proper uses of gardens in general. We have seen the development of an idea which, in consequence, regards private gardens as exhibition grounds, and tender plants of the geranium, verbena, and petunia type as their only proper occupants.

Hibberd was not necessarily against geraniums, as the pelargoniums were known, but there was a time and a place for them. He continued:

The modern flower garden, as known to tens of thousands of persons, is a poor, ginger-bread-entity, ephemeral in respect of its best features, and while demanding but little talent for its production, offering an equally small return in the way of intellectual enjoyment.

Cape pelargoniums and their cultivars bore the brunt of later opinion that:

The stereotyped repetition of scarlet geraniums and yellow calceolarias is in the last degree vulgar and tasteless, and the common dispositions of red, white, and blue are better adapted to delight savages, than represent the artistic status of a civilized people.

Be this as it may, pelargoniums and lobelias are given high prominence in Hibberd's design for parterres and their planting and he covers them in some detail later in his book. It is also reasonable to say that, given that there were so many plants to choose from, pelargoniums held their own and punched well above their weight in the bedding department.

Bedding-out Plants

If Cape plants indoors were going out, as it were, then outdoor ones, or a few of them at least, were reaching dizzy heights of fashion, a phenomenon that can be ascribed to Hibberd's pet hate, the Victorian craze for 'bedding out'.

Bedding-out may be interpreted as an evolutionary step by certain tender plants. Like the air-breathing fish that first poked its head self-consciously above the water and after a few million years grew some legs and left its primeval puddle to explore the dry land, so bedding plants spent the first century of two of their development teetering on the frontier between greenhouse and garden. Philip Miller, for one, had wings of his greenhouse that were partially enclosed in which 'to place many of the most Exotic Plants, which will bear to be exposed in the Summer-season.' Plants from warmer climes such as the Cape were, therefore, making a few first tentative steps into the great outdoors of England.

The Victorian obsession with order and regularity, from empire to household, invaded the garden. There was, it would seem, something of a Chinese influence at work here, gardeners of that part of the world being inclined not to strew their plants haphazardly about the premises but, according to Sir William Chambers 'to dispose of them in great circumspection along the skirts of the plantations, or other places where plants are to be introduced.' Later, Robert Thompson, author of *The Gardener's Assistant*, published in 1859, described how the beds of short-lived annuals from overseas (mainly North America) that had replaced hardy herbaceous plants, did not provide enough enduring colour to see the garden through the summer and autumn.

Filling up the blanks which resulted, was felt to be inconvenient, and this led to the propagation of certain plants which, though requiring protection in winter, were adapted for planting out in the open air in summer. These are termed bedding-out plants.

This strategy allowed for what Loudon called 'The changeable flower garden' whereby certain plants were grown in pots and brought on in the greenhouse, then planted out in the borders at an even age so that they all flowered simultaneously, and were finally replaced with another batch of the same or a different variety when they started to go over.

The origins of this may well have been Oriental, as Chambers suggested, but there seems no doubt that Joseph Paxton was the major mover-and-shaker in the bedding department and that there were other international influences at work. One commentator noted that:

He used all the old patterns of Italy and France for designs of beds, filling them, as had never been done before, with cuttings of tender exotics, which were kept under glass during the whole winter.

In an interesting reflection of the social calendar of the day, Paxton ensured that the beds and terraces of Chatsworth were 'a blaze of colour during the months of August and September' this being the time when the Duke came down from London where he had spent the season.

Pelargoniums and other bedding plants were arranged with military precision in regimented rows and blocks with ne'er a nod towards a natural order of things. This rigidly defined organisation obviously pleased the control freaks, latent or otherwise, of stately homes and city halls, but wasn't to everybody's taste. The early nineteenth century horticultural writer Henry Phillips exclaimed that:

bad taste is seldom more conspicuous than we see trees or plants marshalled in regular order and at equal distances, like beaux and belles standing up for a quadrille or country dance.

Like it or not, a gardening fashion was born and the classic, stereotypical Victorian bedding plants, the beaux and belles so deplored by Phillips, are still popular in some quarters (mainly municipal) today, their major constituents often being pelargoniums and lobelias. Certainly they reached their height in about the 1860s when, as a writer to *The Times* observed, 'convention still ruled the garden with a rod of iron; our fathers were still in the clutches of geometrical formality and hide-bound tradition.'

Despite their popularity in the early nineteenth century, Cape plants gradually fell out of fashion as increasing numbers of new and exciting species were imported from other parts of the world. A further grievous blow was delivered by the arrival of the Wardian Case. A serendipitous invention by Nathaniel Bagshaw Ward (1791–1868), it allowed tender, tropical plants to be transported alive over great distances and to be grown successfully in Victorian drawing rooms. In 1842, Ward wrote a small book entitled *On Growth of Plants in Closely Glazed Cases*, from which this illustrations is taken, and the world beat a path to his door.

Kidneys and Tadpoles

The revolt against glass and bedding gathered momentum under the banner of William Robinson. At his garden at Gravetye Manor in Sussex he demolished all the greenhouses, nominally because he didn't like their contents, but mainly because he didn't like his gardeners sloping off into the warmth on a drizzly day on the pretext of potting-up when they should have been hard at work in the herbaceous borders.

Robinson called the formality of pelargonium-packed beds 'pastrywork' or 'railway embankment' and delivered his rants in his magazine *The Garden: An Illustrated Weekly Journal of Horticulture In All Its Branches*, to whom a regular contributor and future editor was Gertrude Jekyll.

Private gardens soon adopted the 'natural' look espoused by the influential Robinson and Jekyll. Despite their best efforts to kill them

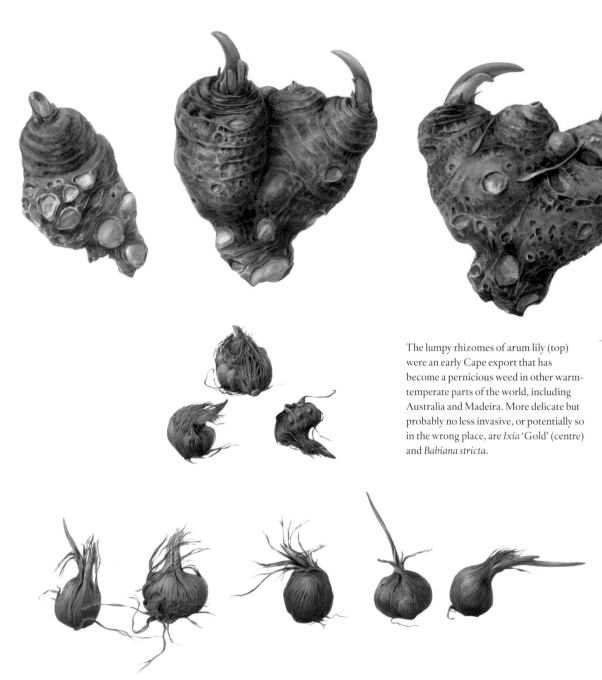

The lumpy rhizomes of arum lily (top) were an early Cape export that has become a pernicious weed in other warm-temperate parts of the world, including Australia and Madeira. More delicate but probably no less invasive, or potentially so in the wrong place, are *Ixia* 'Gold' (centre) and *Babiana stricta*.

To survive in a garden, plants have to be 'garden-fit'. Many plants (for example, fynbos species like ericas, proteas and buchus) have a very sensitive root system which does not like disturbance, as they originate in areas where they did not have to adapt to trampling or grazing distur-bances. Some plants, however, grow easily in any garden, and are able to tolerate human handling, neglect and ill treatment better than others … This ability equips them to withstand the rigours of growing in the garden. So if you blame yourself for the death of your erica, relax, it died for other reasons.

If a plant thrives in gardens within its ecological region I call it 'locally garden-fit' … Most plants with a high 'overall garden-fitness' score originate in the eastern Cape, which can be conferred with the title of 'Cradle of the world's most adaptable garden plants'.

Ernst van Jaarsveld, *Veld & Flora*, 2001.

off, however, bedding plants such as pelargoniums remained popular within certain sectors of society and the 'scores of unmeaning flower beds in the shape of kidneys and tadpoles and sausages and leeches and commas [that] now disfigure the lawn', as one correspondent to *The Gardener's Magazine* had called them 50 years previously, continued to fend off extinction.

Pelargoniums were among the lucky Cape plants. Sooner or later, many others had become casualties of the first greenhouse effect. Through a process of unnatural natural selection, by the end of the nineteenth century those Cape plants that were destined to endure, at least in England, were well established, while the majority had fallen by the wayside, innocent victims of fashion and hot flushes.

One of the most remarkable features in the South African Government Pavilion at Wembley was the weekly display of fresh wild flowers in the Cape Town Section. These cut flowers were transmitted in the cold storage room of the Union Castle steamers and a regular supply arrived every week as if they had been newly picked. The general public showed very great interest in this exhibit of indigenous Cape plants, whose beauty and attractiveness were enhanced by the tasteful arrangement in the vases by Mr Dunsdon, of Caledon, Cape Province.

It is much to be regretted that the cultiva-tion of these remarkable plants has nearly died out in this country, and it is to be hoped that the consignors of the plants to Wembley will continue their good work and help to fill our cool greenhouses with some of the beauti-ful Ericaceae and Proteaceae, which were such a conspicuous feature in our gardens in the early part of the last century.

Bulletin of Miscellaneous Information, 1939.

CHAPTER NINE

The Treasures of this Beautiful Tribe

Set against a background that interweaves expanding trade and botanical exploration with garden fashion and development, the horticultural history of the Cape begins with bulbs and their kin.

Although a dried-up protea was the earliest plant to be collected at the Cape and brought back to Europe, it was bulbs that first made the journey alive and in good enough condition to be planted and to grow in a different hemisphere. Conveniently for the gardener, bulbs can be dug up and transplanted and barely register that they've been moved. As William Hemsley, keeper of the Kew herbarium at the turn of the nineteenth century, put it:

> Bulbous plants and succulents are more tenacious of life than most other portable ones, or even seeds. We find accordingly that a large proportion of the earlier introductions belong to one of these categories.

'Bulb' in its broadest sense describes the underground storage organ that enables a plant to survive difficult times. Plants having such a capacity are called geophytes and there are well over 1,000 species of them in the drought- and fire-prone Cape. When these plants die back after flowering and drought or fire destroys their leaves and other above-ground parts, a subterranean fleshy bulb, corm, rhizome or tuber sees them safely through to the next growing season. Where there is year-round rainfall and protection from fire, notably in the east of the region, some geophytes that grow in sheltered, woody kloofs with perennial streams can afford to be evergreen, but still have their bulbous reserves to fall back on.

Although the Cape can boast a natural abundance of bulbs, only a relatively small number have so far found their way into mainstream cultivation. There have also been more losses from, than additions to, the list over the years, and many remain the preserve of the specialist grower. As long ago as 1852, William Thompson, founder of the nursery that later became Thompson & Morgan, considered that 'It is difficult to resist the conviction, that to a large class of amateurs, the treasures of this beautiful tribe are comparatively unknown.' This is as true today as it was then.

Boiled Chestnuts and Floating Butterflies of Bright Lilac-blue

It has been suggested that the first Cape bulb to be successfully grown in England was a *Moraea*, possibly *M. ciliata*. This diminutive member of the iris family is common on the Peninsula and is reputed to have been brought to England as early as 1587, having been taken on board at the Cape to supplement ships' rations. Accurate identification would have been important because although some corms were eaten by the Khoekhoen and were said to taste like 'boiled chestnuts', others are poisonous to humans and livestock.

This claim of first arrival remains unsubstantiated but there is, on the other hand, no doubt that moraeas were being cultivated in Holland

The Cape Floral region is a botanical anomaly. Not only does it have more plant species than would be anticipated given its latitude and climate, it is also home to far more bulbous plants than anywhere else in the world.

Geophytes occur widely among flowering plants, in families as diverse as the daisies (Asteraceae) and grasses (Poaceae). Among this diversity, however, few species actually form true bulbs, and 'bulb' has been rather loosely used for any geophyte that is more or less horticulturally interesting, with the consequence that the boundaries of the group are rather loosely and idiosyncratically drawn. Peter Goldblatt *et al.* The Colour Encyclopedia of Cape Bulbs, 2002.

opposite
The Scarborough Lily *Cyrtanthus elatus* takes its name from the Yorkshire seaside resort where, in about 1800, it is said that amongst the cast-up cargo of a wrecked Dutch East Indiaman were bulbs of this Cape beauty. Formerly known as *Vallota speciosa*, and sometimes still sold under that name, the Scarborough lily is evergreen and one of the most rewarding Cape bulbs. On its home turf it is known as the George or Knysna lily, after the towns near which it occurs.

(where *M. fugax* was described as 'an old denizen of Dutch gardens') and England by the mid-eighteenth century.

The genus was christened by Philip Miller in 1758. He gave it the name of *Morea* [*sic*], explaining that:

> I have taken the liberty of constituting this Genus of Plants and applying this title to it, in Honour of Robert More, esquire, of Shropshire, who is a very learned Gentleman and well acquainted with more Branches of Natural History, particularly with Botany.

Linnaeus later accepted this name but transcribed it as *Moraea*. While there is a suspicion that he might have misappropriated the name and tweaked it to honour his wife's family, Moraeus, it may have been a genuine typographical error.

Moraeas featured regularly in *Curtis's Botanical Magazine* for a few years before slipping elegantly into virtual oblivion as far as adorning English gardens, greenhouses and windowsills is concerned. This is a shame, as a prettier group of plants would be hard to imagine. Perhaps they are handicapped by having flowers that, in some species, last only a single day (although there is a regular succession of them), fading by nightfall, and a flowering season lasting only a few weeks.

This floral ephemerality should not, however, prejudice the potential cultivator, as plenty of popular plants display these traits (*Hemerocallis* are, after all, called 'day lilies' for a reason). With their flowers of white, yellow, or delicate shades of blues or purple, moraeas certainly make as attractive a pot plant as, for example, miniature irises. Those bred from *M. villosa* with 'peacock' centres to their flowers are strikingly beautiful, and it seems odd that they have not been developed and popularised indoors in cooler climes as they have been outdoors in Australia and the warmer parts of North America.

Writing in the *Journal of the American Horticultural Society* in 1948, Claude Hope waxes lyrical about *M. polystachya*, one of the larger and more floriferous species, but I'm sure he would have applied his words to many of its 115 or so compatriots.

> It seems hopeless to attempt to convey in words the beauty and loveliness, the fragile charm, the cheery colours of the flower; only an artist's brush could do it justice. Like floating butterflies of bright lilac-blue, decorate with yellow, the flowers rest lightly above the long slender spathe valves, five or even more on each slender inflorescence. Reminiscent of some of the beardless irises, but more graceful and more generous of flowers than most, a group of these is an exciting vision.

Although the appeal of moraeas remains mysteriously limited, two plants that have their origins, if not their entire horticultural history, rooted at the Cape have become enduring favourites worldwide: *Gladiolus* and *Freesia*.

The Sword Lily

The gladious in all its multifarious gaudiness is one of the world's most popular garden and cut-flowers. Many of today's varieties display

above
Cape bulbs made a very favourable early impression on the gardeners of Europe. This *Gladiolus* by Georg Dionysus Ehret (1708–1770) featured in *Plantae selectae* of Christopher Jacob Trew (1695–1769), a Nuremberg physician.

opposite
Moraea villosa
'Villosa' means hairy, although the slight hairiness of the leaves and stems is hardly the most striking feature of the species. Occurring in a variety of soil-types in the south-western Cape, it grows up to 40 cm and has relatively large and long-lasting blooms in spring. Insect damage and over-watering are the main obstacles to its success in cultivation.

relatively little of their refined Cape ancestry, however, it having been largely swamped by assertive brethren from the north of the subcontinent and a history of cultivation which, to some tastes, sought brashness over elegance and ease of cultivation rather than a horticultural challenge.

Gladiolus is the diminutive of *gladius*, the Latin for a sword (as used by gladiators), and refers to the shape of the leaves. There are six species in Europe while sub-Saharan Africa is home to some 240, of which the Cape Floral Kingdom holds 105 (86 of them endemic). This plenitude has been ascribed to an evolutionary history under which the plants competed for, and adapted to, a diversity of pollinators, such as sunbirds, bees, and various flies and moths with long tongues.

The first Cape *Gladiolus* to make an appearance in Europe was the orange-and-greenish-flowered *G. alatus*, sent to Holland in the late seventeenth century. Material and illustrations (more often than not copied from one publication to another) also found their way to England and, by 1691, this particular plant was figured in Plukenet's *Phytographia*. A second species, *G. angustus*, which has a very fine flower of cream to pale pink, diamond-marked with red, featured in Boerhaave's 1727 account of the plants growing in the Leiden gardens.

Almost 20 years later, the night-scented *G. tristis* was blooming in the Chelsea Physic Garden, grown from seed that Philip Miller had received from Dr Job Baster in Holland. By 1758, Miller could boast at least four Cape species, and the following year *G. tristis* was also flowering in the university garden at Uppsala, Miller being the most likely source of seeds or bulbs.

In 1790 *G. cardinalis* featured in the youthful *Curtis's Botanical Magazine*:

Gladiolus cardinalis

Plate 135 from *Curtis's Botanical Magazine*, 1791 '… a very strong plant of it flowered also this summer at Messrs. GRIMWOODS and Co. which divided at top into three branches, from one of which our figure was drawn.'

This new species of Gladiolus of whose magnificence our figure can exhibit but an imperfect idea, was introduced into this country from Holland, a few years since, by MR. GRAFTER, at present gardener to the King of Naples … It is most probably a native of the Cape, flowers with us in July and August, and is increased by offsets from the bulbs; must be treated like the ixias and other similar Cape plants.

A later enthusiast declared: 'What flower can surpass the brilliancy of the *G. cardinalis* when grown in luxuriance? Here, like the sun in the planetary system, it is the centre of attraction.'

Following this dazzling but limited entrance onto the horticultural stage, a wider range of gladioli found themselves among the many Cape species that became hugely popular in the 50 years spanning the last decade of the eighteenth century and the first four of the nineteenth. It was in the early years of this period that the enthusiasts started seriously tinkering with their 'glads'.

As with many of the pioneering experiments in plant breeding, the earliest recorded hybridisation of gladioli was undertaken by an amateur, in this case William Herbert. Writing in the *Transactions of the Horticultural Society* in 1820, Herbert describes the products of hybrids involving *G. angustus*, *G. carinatus*, *G. carneus*, *G. caryophyllaceus*, *G. liliaceus* and *G. tristis*. The resulting crosses between late spring- and early summer-flowering species were sterile but could, he reported, grow permanently 'on the light yellow loam suitable for barley' in his Yorkshire parish, and 'flowered every June among the roses.' In 1837 he reported that some clumps had persisted for over 20 years.

In another article Herbert declares:

I am persuaded that the African gladioli will become great favourites with florists when their beauty in the open border, the facility of their culture and the endless variety which may be produced from seed by blending the several species are fully known, nor will they be found to yield in beauty to the Tulip and the Ranunculus [anemone].

Even he could not have imagined just how much the industry that he essentially set into motion, would grow.

A sterile hybrid called 'G. pudibundus', the 'Blush-flowered Corn-flag', featured in Robert Sweet's *British Flower Garden* (Volume 2, 1833). The accompanying text describes it as 'a hybrid offspring, we believe, between *Gladiolus cardinalis* and *blandus* … raised by the Honourable and Rev. William Herbert. It is an extremely showy plant, and therefore cannot fail to become a universal favourite with florists'. *G. blandus* is now known as *G. carneus* and shares the popular name 'painted lady' with several other Cape gladioli. In 1823 the first hybrid with a direct link to today's gladdie cornucopia was bred by Colvill's Nursery in Chelsea; a richly-scented, yellow-blotched scarlet beauty, it owed its colour to *G. cardinalis* and its fragrance to *G. tristis* var. *concolor*. This also appeared in Sweet's book, and he called it 'G. colvillei' (a name frequently misspelt over the years).

Two further hybrids – 'G. ramosus' and 'G. insignis', played an important part in the development of the garden gladiolus. The former

'Early Flowering Gladioli'
from *The Garden*, 1888.
'Blushing bride (fig. 1) is a very pretty flower with a white ground and dark crimson spots on the lower petals, the blotch in the central one being large and broad. Rosy gem (fig. 2) is a flower of a bright rose colour with a darker blotch of the same; very free and strong flowering. General Scott (fig. 3) has a white ground with a yellowish blotch margined with bright crimson, and is a distinct and pretty flower. *Rosea maculata* (fig. 4) is larger flower of a brilliant scarlet-rose colour with light blotches on the lower petals, margined with crimson; very showy.'

I have found, with regard to the more vigorous-growing Gladioli, that a sprinkling of well-decomposed cow-dung over the drainage conduces much to the strength of the plants. At the same time, I prepare a bed of old dry tan mixed with some fresh stable manure, at least two feet in thickness above the level of the ground, and place on the top a large frame; within this I plunge the pots in the old tan, the stronger and taller growing Gladioli at the back, Tritonias, Sparaxises, Ixias, Babianas etc, arranged in due gradation in front, down to the dwarf Oxalises and Lachenalias. At night I put on the lights, and in fine weather give abundance of air … The natural order of Iridaceæ is very extensive, and of Cape species, I know none which will not be found to bloom in perfection under my plan of treatment, and which I have endeavoured to make intelligible to your readers, and shall be gratified if I succeed in eliciting any communications thorough your medium on my favourite tribe.

'C. B. S.', *The Floricultural Cabinet*, 1826.

is believed to be a hybrid of *G. cardinalis* and *G. oppositiflorus*, while *G. insignis* sprang mysteriously from Colvill's collection in 1839 after it had been sold to Messrs Lucombe, Prince & Co. of Exeter. A deep red flower with violet markings, its parentage is unclear although one source claims *G. carneus* as making an input.

In the twentieth century, the eminent Dutch horticulturist Ernst Krelage back-crossed '*G. ramosus*' with *G. carneus* or *G. cardinalis* to produce dwarf, early-flowering hybrids, known today as '*Ramosus*'. These showed many of the attractive features of the wild Cape parents, particularly the elongated blotches on the lower tepals. When back-crossed with *G. cardinalis*, these produced another lineage, the '*Nanus*' (dwarf) hybrids.

Further hybrids of '*Colvillei*' and '*Nanus*' were subsequently developed and, as relatively delicate, spring-flowering cultivars, they were very popular for a few years before being bullied out of fashion by monstrous glads from continental Europe bred predominantly from new species introduced from further north and east in southern Africa. The delicate Cape cultivars are, however, now making a welcome comeback.

There are well over 10,000 named cultivars of *Gladiolus* today and, to my mind, they are at their most attractive when they bear the strongest resemblance to their Cape ancestors. Many Cape species are highly scented, so it would be wonderful if some of their perfume could be reintroduced to make a garden- or pot-plant that not only looks but smells exquisite.

Destitute of all Spotting

A number of Cape bulbs that enjoyed taxonomic independence in the past have now been transferred into the genus *Gladiolus*. One that has successfully resisted this move, however, is *Freesia*, although when it was first discovered it was included in the genus *Ixia* before being shifted temporarily into *Gladiolus*. In 1759, two Cape *Freesia* species flowered for the first time in Europe and specimens of them are in Burman's herbarium. He called them *Ixia corymbosa* and *I. caryophyllacea*.

In 1786, Jacquin described another new *Freesia*, naming it *Gladiolus*

It is incongruous that we in South Africa now import gladioli corms by the million from Europe, thus buying back some of the floral wealth that once grew wild on our veld. Holland, Britain and the U.S.A. are leaders in the production of new varieties of gladioli. At the Wisley trial in England [in 1959] 127 new varieties were grown under test conditions and out of the twenty-one awards made by the Royal Horticultural Society no fewer than seventeen went to gladioli raised by gardeners of Noordwyk, Holland. Only one variety bore a name with a South African association – it was 'Cecil Rhodes' ... Australia was represented with a solitary award named 'Janet Smith'; but South Africa, home of the species on which the gladiolus hybridization industry of the world has been founded, was not represented at all.

Conrad Lighton, *Cape Floral Kingdom*, 1960.

left
Of the thousands of gladiolus cultivars developed over the years, the 'Nanus' varieties retain most characteristics of the original Cape stock. This is 'Nanus Claudia'.

right
When they came into bloom in our garden, two of these gladiolus cultivars bore rather little resemblance to the picture on the packet in which they arrived. The third, 'Stella', proved to be quite variable in the extent and distribution of red on the tepals.

above
Freesia alba
A field sketch made on the southern Cape coast. What a painting cannot convey is the exquisite scent of this delectable species.

opposite
Freesia cultivars
From the rather limited palette of the wild species, myriad freesia cultivars have been developed. It is now one the most popular cut flowers in the world, with the Dutch nurseries alone producing in the region of half-a-billion stems each year.

refractus, and others were later placed alongside it in the genus *Gladiolus*. It was not until 1827 that Christian Ecklon (1795–1868) published a generic name *Freesea* [*sic*] after his friend Friedrich Freese, a physician at Kiel, and in 1866 taxonomic chaos finally gave way to order and the various species were placed together in the genus *Freesia* by the German botanist Friedrich Klatt.

Freesias only came to the attention of horticulturists, however, after 1874, in which year Max Leichtlin, the distinguished botanist and plant collector of Baden-Baden, came across a yellow species 'among some neglected plants' in the botanic gardens at Padua in Italy. He propagated plants from this stock and the species, originally from the sandy coastal strip east of Cape Agulhas and subsequently named *F. leichtlinii*, became very popular with gardeners. The introduction in 1878 of the wonderfully scented *F. alba*, also a southern Cape coastal species, bolstered the reputation of *Freesia*.

Meanwhile, an overlooked colour variety of *F. corymbosa*, originally known as *F. armstrongii* after its nominal discoverer, had been flowering merrily at Kew as long ago as 1826. Although William Armstrong had collected corms of this plant in the southern Cape in 1896, material had been sent from there to Kew by James Bowie many years before but the plant's potential had not been recognised. The Armstrong variant, however, with its unique pink to red flowers, was to form an important constituent of cultivars, as noted by a correspondent to the American magazine *Horticulture* in 1909:

In the year 1901 the [Dutch] firm of Van Tubergen received from an English correspondent, Mr. Armstrong, resident in S. Africa, several freesia tubers with the remark that they were pink flowering. Naturally Herrn Tubergen was somewhat suspicious concerning the colour of the newcomer, as all the freesias obtained from that part of the world were outwardly of a brown tint, or white with yellow staining…

Having proven its colourful credentials at van Tubergen's nursery, '*Freesia armstrongii*' was crossed with '*F. refracta alba*' to produce a floriferous hybrid. The most highly coloured were then back-crossed with the 'largest flowered *F. refracta alba*, snow white and destitute of all spotting …' to produce '… seedlings … of carmine and rose tints, with many gradations.'

By the mid-twentieth century, about 50 named freesia cultivars were available in a wide variety of colours, although only a few white or yellow ones retained a scent of any potency. It seems that as the colours strengthened and diversified, the perfume that makes freesias so attractive was largely lost. Recent research has revealed that up to 16 volatile compounds give the scented freesias their gorgeous sweet or lightly-spicy fragrance, of which linalool and terpinolene are among

Agapanthus africanus

A relatively low-growing species, *Agapanthus africanus* is found in montane areas of the south-western Cape. It is very attractive but unfortunately quite difficult to cultivate.

the most important. A major challenge to breeders is to prevent the loss of this defining virtue.

Freesias may claim to be the most popular cut flower in the world today, with the bulk of the tens of millions of stems sold annually in the UK being imported from the industrious Dutch nurseries which produce an estimated 500 million a year. An impressive result for a humble Cape bulb and a chance encounter in an Italian garden.

The Lily of the Nile

One of the disconcertingly familiar flowers that first caught our untrained eye in fynbos when we arrived in South Africa, *Agapanthus* is another Cape speciality that has been spread far and wide to the gardens of the world.

In 1679 an agapanthus was described in glowing terms in Jacob Breyne's *Prodromus Fasciculi Rariorum Plantarum*:

> This African Hyacinth, which in the month of September last year flowered in the garden of the most illustrious and honourable Lord Hieryonymius Beverninck, is a remarkable wonder among the rare plants of Europe, crowned with flowers each of which is singularly beautiful.

The illustrations to accompany the *Prodromus* did not appear until 1739, but they included what was clearly a species of *Agapanthus*.

Meanwhile, in 1687, *Agapanthus* was mentioned in Paul Hermann's catalogue of the Leiden Botanic Gardens. By the end of that century it had also flowered at Hampton Court Palace, being described by Plukenet as *Hyancintho affinis, tuberosa radice, Africana, umbella caerula inodora* which, as the 'Tuberous African hyacinth with umbels of scentless blue flowers' sums it up pretty well. This particular plant was probably derived from the Dutch stock recorded by Breyne and may have accompanied William and Mary when they relocated to England in the Glorious Revolution of 1688.

The plant was named *Crinum africanum* by Linnaeus in 1753, but in 1788 it was recognised as being too distantly related to *Crinum* to bed down in this genus and so was placed in *Agapanthus*, created for it by Charles L'Héritier. The generic name means 'love flower' and was applied for no obvious reason. He named the species *A. umbellatus*, an umbel being a flower cluster, from the Latin *umbella*, a sunshade. Strictly speaking, it should have had the specific epithet *africanus* in accordance with the original Linnaean name.

Sometime around 1890 the deciduous species *A. campanulatus* was introduced from north-eastern South Africa. It was originally known as *A. umbellatus* var. *mooreanus* after David Moore and his son Frederick, to whom the original material was sent at the Glasnevin Botanic Garden in Dublin. Crossed with existing Cape stock, its hybrid descendants are often sweepingly termed '*A. umbellatus*' which, despite more recent reclassification, remains the inaccurate but convenient name of a whole range of cultivars.

Significant advances in the development of *Agapanthus* were made in the 1940s by Lewis Palmer who, using seed from Kirstenbosch

National Botanical Garden, Cape Town, developed the 'Headbourne Hybrids', named after his property in Hampshire; these were, it has to be said, probably derived largely from the non-Cape species *A. campanulatus*. In the 1970s, Palmer's plants formed the major component of a trial of *Agapanthus*, designed to assess the merits of different cultivars, planted at the Royal Horticultural Society's garden at Wisley. Many of these Wisley plants went on to be trialled in the Netherlands in 1978 and, by 2004, Wim Snoeijer was able to list well over 600 cultivars in his revision of the genus.

This is certainly not the end of the story, and the 'Lily of the Nile' (one of the plant's many colloquial misnomers) remains one of South Africa's most successful floral exports.

The Sea Nymph

In the early 1770s Francis Masson collected some bulbs from a rocky Cape mountain slope and packed them off in one of his many Kew-bound consignments. When the bulbs flowered it became apparent that it was the same species that had, for many years, been thought to have come from Japan via the Channel Isles. The latter place and a slice of historical whimsy even gave the plant its name – *Nerine sarniensis*.

As long ago as 1634 this particular flower, with its glittering, gold-dusted 'petals', had burst into bloom in the Parisian garden of botanist and physician Jean Morin. Even the day (7 October) is recorded and an illustration of it published by Jacob Cornut the following year was accompanied by the description '*Narcissus japonicus rutilo flore*' – 'the reddish-orange flowered narcissus of Japan'. On the basis that it had been included in a batch of bulbs and other material sent from the Far East, but added to when the ship called in at the Cape of Good Hope on the return journey, this is an understandable error. Or it may be that Morin and Cornut actually mistook it for a genuine Japanese species, the red spider lily *Lycoris radiata*, which it resembles.

Some 20 years later the plant was found growing on Guernsey in the Channel Islands. An explanation attributed to Robert Morison, Professor of Botany at Oxford, proposed that a Dutch ship returning from the Far East had foundered near the island and among its cargo had been some bulbs that were cast high up the beach and were able to establish. An alternative of this version of events is that the bulbs were presented to an hospitable islander by a survivor of the shipwreck. Either way, the plants found the mild, sandy conditions to their liking and soon settled in.

The species was later named *Amaryllis sarniensis* by Linnaeus, the epithet being derived from *Sarnia*, the Romans' name for Guernsey, and so the plant's supposed provenance was entrenched. Nerines were more likely, however, to have been introduced deliberately to the island, their agent being Major General John Lambert, a keen plantsman who had obtained stock, probably from Morin, and grown them in his Wimbledon garden. A specimen painted by Alexander Marshal (*c.* 1620–1682) has the inscription 'this flower was sent me by Generall Lambert august 29 1659 from Wimbleton.'

Lambert, a Parliamentarian, was exiled to Guernsey in 1662 following the Restoration, apparently taking with him his precious nerines and other bulbs (he was a tulip fanatic and known disparagingly in some quarters as 'Knight of the Golden Tulip') and continuing to garden enthusiastically on his temperate island outpost.

Whatever the nerine's true horticultural origins, a century-and-a-half later Dean Herbert had the enigmatic bulb in his collection. In his revision of the amaryllid family he removed it from *Amaryllis* and created for it a new genus, *Nerine*, named after one of the Nereid sea nymphs. These were the 50 daughters of Nereus, god of the Aegean, and his wife Doris. Helpers of mariners in distress, the nereids are traditionally depicted in classical renderings astride dolphins or large fish, and in neoclassical paintings wearing as little as possible. The watery history of *Nerine sarniensis*, if ever it had one one, is thus commemorated.

Masson's subsequent discovery of the self-same plant growing in the mountains of the south-western Cape was enough to enlighten botanists as to its true birthplace, but by now the Channel Islanders had embraced it as their own, to the extent that it even had the common name of Guernsey lily. It has since featured on the islands' postage stamps and remains an important contributor to the local cut-flower industry, although the careless application of the name to other *Nerine* cultivars and species grown there makes it unclear as to what exactly you will receive if you order a bunch of 'Guernsey Lilies'.

The 'wild' nerines of Guernsey and those that found their way into cultivation vary in colour. Those flowering in Paris in 1634 were described as being 'something near to that of Cinnabar, or the finest sort of Gum lacca', while English ones of 1725 were 'like a fine gold

Just as the bluebell grows in England, casting a sheet of dark blue upon the earth, so the agapanthus grows round [the south coast town of] George. And that stately flower is no delicate stranger which has to be hurried into the greenhouse with the approach of winter: it stands over three feet high from its native soil, bearing great flowers of blue or white, which you can just about cup in both hands. I have grown agapanthus in England but, having seen it round George, I shall never trouble to do so again.

H. V. Morton, *In Search of South Africa*, 1948.

Tissue, wrought on a Rose-colour'd Ground' that then faded pink. A classic scarlet bloom was later portrayed thus:

> Their Colour is the most perfect Red, and they are spangled all over as it were with Gold; this with a deep red vein running along their middle, gives them a Glory and a Splendour when viewed in the Sun, superior to that of all other Plants.

The observer goes on to comment:

> Nature, not to lavish her treasures upon one Flower, has denied this Fragrance. One sense is fully satisfied with it, and even more, for in the Sun the Eyes ache to look upon it.

Of the c.23 species of *Nerine*, four (*N. sarneinsis, N. humilis, N. pudica,* and *N. ridleyi*) occur in the Cape Floral Kingdom, but only *N. sarniensis* has made any significant contribution to floriculture. A greater input to the cultivars has come from species found north and east of the region, in particular the pale pink, frost-hardy *N. bowdenii*.

As early as 1837 Herbert had created nine hybrids involving crosses between three species, including *N. sarniensis,* and an increase in their popularity followed in the late nineteenth and early twentieth centuries. Henry Elwes of Gloucestershire succeeded in expanding the colour range through his crosses, many of which, after his death in 1922, were inherited by the Stephenson Clarke family of West Sussex who went on to produce a range of hybrids named after their garden, Borde Hill.

A little later, Lionel de Rothschild began to produce what became the great range of Exbury Hybrids. Although his collection was sold in 1974, his grandsons restored the family interest in the plants. Many of those that they use in an extensive programme of breeding and selection are descended from original Exbury stock gifted back to the nursery by Sir Peter Smithers, who had bought part of the original collection. Smithers, one time Member of Parliament for Winchester, succeeded in creating over 1,000 nerine crosses and, when assessing their colourful qualities, traditionally displayed them to best effect under spotlights against a sapphire-blue backdrop!

One nineteenth-century aficionado, James O'Brien, observed:

> Too often these beautiful bulbs are ruined with kindness ... The chief points to be observed in their management is to give them a long and decided period of rest by drying them off ... not a drop of water should be given them until the [flower] spikes appear...The beauty of these Guernsey Lilies, their easy culture, and the long duration of their flowers should make them general favourites with amateurs and window gardeners.

Guernsey Lily

Nerine sarniensis from *The Garden* of 1882. 'The great beauty that these plants possess, and the admiration they command wherever they are well bloomed, make one wonder that they are not more generally met with than they are.' *RHS, Lindley Library*

Little Weeds and Harlequins

While the history of horticultural development for plants such as *Agapanthus, Freesia, Gladiolus,* and *Nerine* have been relatively well documented, other Cape bulbs seem to have just appeared on the scene without much fanfare and rather little subsequent commentary. One such is *Ixia,* known as African corn lilies everywhere apart from in Africa. Fifty species occur in the Cape, every one of them beautiful and under-appreciated in about equal amounts. They are also, according to one Victorian gardener:

> Cheap, easily grown, and when in flower full of elegance and rich in brilliant colours...These little Cape weeds should, and indeed do, find wide favour with us as pot plants for spring flowering and for the beautifying of warm, sheltered borders out-of-doors in summer. For a few shillings, one may procure almost a peck of Ixia bulbs in 'fifty or a hundred of the finest and newest varieties' – good, sound bulbs, too, such as will not fail to flower if their requirements, which are simple, are afforded them.

It is likely that all the cultivars were descended from a small stock of *Ixia* species, including *I. maculata, I. monadelpha* and *I. patens,* introduced in the last two decades of the eighteenth century, mainly by Francis Masson. Given the range of colours and their ease of cultivation, it's something of a mystery as to why there are so relatively few varieties available today compared to the nineteenth century when they were clearly all the rage.

In 1873, Shirley Hibberd was at pains to point out that 'The Continental growers have been outstripped by the English in the cultivation of these most elegant flowers.' In this context he was referring not only to ixias, but to a close relative and another member of the iris family, *Sparaxis,* the harlequin flower. A particularly striking species, *S. grandiflora,* was grown at Kew as early as 1758 and at Leiden in 1776, but it was not for another 150 years that the most popular and enduring of the *Sparaxis* hybrids was developed, and these arose almost by accident.

All the sorts [of ixias] multiply very fast by offsets, so that when once obtained, there will be no occasion to raise them from seeds: for the roots put out offsets in great plenty, most of which will flower the following season, whereas those from seeds will not thrive through the winter in the full ground in England, so should be planted in pots, and placed under a frame in winter, where they may be protected from frost, but in mild weather should enjoy the free air; but they must be guarded from mice, who are very fond of their roots, and if not prevented will devour them.

Philip Miller *Gardener's and Florist's Dictionary,* 1724.

A selection of ixias depicted in *The Garden* magazine of 1884. The accompanying text is lavish in its praise of the flowers' looks, but is pragmatic about their prospects: 'In England the cultivation of these plants out of doors does not meet with much favour, owing to the unfitness of an average English season for their growth and the production of flowers. In a few nurseries, however, and in some private gardens in the south, very fair success has been met with in the out-of-door management of Ixias.'
RHS, Lindley Library

Only a few of the 14 species of *Sparaxis* in the Cape Floral Kingdom have been brought into cultivation. Two of these, *S. elegans* from the north-western edge of the region and *S. grandiflora* from the Tulbagh Valley, were planted by Japie Krige, a pioneer grower of indigenous wildflowers, in his Stellenbosch garden in the early 1930s. The former species is salmon-pink in colour, occasionally white, and with a purple-and-yellow centre; the latter is a deep plum. Finding the miles that separated their natural populations reduced to a petal's distance, the two crossed readily. Their progeny provided a startling spring display of multicoloured blooms until, after a few years, the whole lot succumbed to a fungal infection. By this time, however, Japie's garden had been visited by thousands of admiring visitors. He also generously gave seed to anyone who asked for it.

Japie Krige's home-grown harlequin hybrids have provided the material and stimulus for further crossing and today we can get a combination of cream, red, purple and yellow and combinations thereof from a single packet of *Sparaxis* bulbs, one of the most colourful products of the Cape Floral Kingdom.

Homer's Cicada and Dove's Dung

The idea of shipping vast numbers of cut flowers from South Africa to England in the nineteenth country sounds like one of those schemes in which you would be disinclined to invest your life savings. Nevertheless, such was the resilience of the blooms in question that this is just what was done and, until the same flowers began to be cultivated in Europe, they supported quite a thriving seasonal industry around Cape Town.

The flowers were chincherinchees *Ornithogalum thyrsoides*, known locally by their affectionate diminutive of 'chinks'. Members of the genus are also found in eastern Europe and the Middle East, where they are known as Star of Bethlehem. The name *Ornithogalum* means 'bird's milk'. The origin of this terminology is obscure; attempts to explain it have included the whiteness of the flowers, juice from the stems, or the scattering of the white flowers over the landscape as if deposited from on high by overflying birds. The Middle Eastern species *O. umbellatum* is referred to in the Bible as 'Dove's Dung'.

The South African vernacular name chincherinchee was applied as early as the eighteenth century and is an onomatopoeia, referring to the squeaky sound made if the stems are rubbed together. A line in Homer refers to the 'lily-like sound of the cicada', a phrase that can most convincingly be explained by invoking the squeaky stems of the local ornithogalums.

The first chink had made its way from the Cape to Europe long before a cut-flower industry was established. In his *Curae Posteriores* of 1611, Clusius described what is thought most likely to have been a species called *O. conicum*, writing:

> Bulbs of the genus Ornithogalum were brought to Amsterdam by Dutch sailors, who had collected them where they grew, in some bay situated to the west of that extreme and celebrated Promontory of Aethiopia

CAPE BULBS *are remarkable for the beauty of their flowers; and as they occupy but little space, a considerable collection of them may be grown in a very small garden, in a great measure without the aid of glass.*

Mrs Loudon, *The Ladies' Companion to the Flower Garden*, 1849.

above & below
Jane Webb Loudon (1807–1858)
wife of John Loudon, was herself a talented writer and became an accomplished botanical artist. The diversity and beauty of the Cape bulbs cultivated at the time are depicted in her paintings in *The Ladies Garden of Ornamental Bulbous Plants*, published in 1841.

above
Wild *Sparaxis*
The striking colours and markings of wild *Sparaxis* species combine to produce a wide variety of shades and patterns among today's cultivars.

above
***Sparaxis stellaris* (now *grandiflora*)**
from Robert Sweet's *The British Flower Garden* (1827). This species is one of the original parents of the modern 'harlequin' sparaxis hybrids.

above

Wild chincherninchees *Ornithogalum thyrsoides* were once gathered in their thousands at the Cape and sent in bunches by ship to England to brighten winter households. They are now crossed and cultivated on an industrial scale in Europe to sell as bulbs or, like these cultivars from a Scottish florist, cut flowers.

below

Arum lily frogs

secrete themselves by day within the chalices of their eponymous plants. This pretty, if gangly, frog is widely distributed along the southern Cape coast but is vulnerable to habitat loss through urbanisation and agriculture.

above

Arum lily *Zantedeschia aethiopica*

by Georg Ehret from *Phytanthoza iconographia*, the florilegium of Johann Wilhelm Weinmann (1683–1741). Most modern Cape bulbs in cultivation have resulted from careful selection and breeding, but one species that has needed no improvement is the arum lily, often known overseas as the calla lily. It grows like a weed in marshy areas at the Cape where it is called 'varkblom', the pig flower, because pigs eat the fleshy rhizomes. '*Zantedeschia*' celebrates Italian botanist Giovanni Zantedeschi (1773–1846) while '*aethiopica*' is shared by many plants from the Cape, broadly referred to in its pre-colonial infancy as Ethiopia or Africa.

Ornithogalum dubium
a widespread species at the Cape,
has given its intense orange colour to
a range of modern cultivars. Crossing
with *O. thyrsoides* produces a sturdy,
floriferous hybrid that can bloom for
several months.

commonly called the Cape of Good Hope, where they had rested for
some days to refresh. At the end of August 1605, Christian Porrett
brought from Amsterdam a full stem of flowers of Ornithogalum which
he gave to me and from which and from a few other characters of which
I learned, I have put together this account.

The plant was subsequently described officially and illustrated by
Jacquin who grew a number of Cape chincherinchees at Schönbrunn.

In the nineteenth century, white chincherinchees in bud were
collected by the armful in mid-summer on the Cape west coast, where
they grow in profusion on damp, sandy flats. Suitably packaged, they
were despatched to England and, despite enduring weeks out of water
and out at sea, the flower buds would open to brighten gloomy Victorian
parlours and drawing rooms in the winter.

The historian Conrad Lighton tells how this industry was established
by Hildagonda Duckitt. Visitors to her family farm, Groot Poste, in the
late 1870s included the daughters of the Governor of the Cape, Sir
Henry Bartle Frere. The girls had a fondness for the chincherinchees
and, knowing of the flower's durability, the Duckitts sent them bunches
when they returned to London in 1880. Once at their destination in
time for Christmas and popped into a vase, the lower buds would be
the first to open, the topmost ones lasting until Easter.

Their popularity established by the enthusiasm of the Freres and the
initiative of the Duckitts, chinks were soon grown in the northern
hemisphere to satisfy the growing European market. The demand for
the genuine Cape article nevertheless remained strong. In the 1950s,
while an English nursery was selling bulbs for three shillings and
ninepence a dozen or 7/6 for 25 (postage and packing 1/-), cut flowers
shipped from Cape Town to the United States cost $3.50 for 25 stems.
For 2/6, Capetonians could send a shoebox full, the cut stems sealed
with candle wax, to friends and family in Europe. In a record year some
55,000 boxes of 50 stems apiece were exported commercially. All these
were harvested from the wild, and such was the value of the crop that
flower poaching became a serious problem.

White chinks of one variety or another, largely a product of the
Dutch breeders, are widely available today, both as bulbs and cut
flowers. Another species, the striking yellow *O. dubium* that occurs
almost throughout the Cape Floral Kingdom, is similarly ubiquitous
overseas. A number of hybrids have been developed that combine the
best features of the African and Eurasian species for the cut-flower and
bulb markets. So, 150 years after being fondly picked and packed and
sent to London from the Cape, the lily that squeaks continues to
brighten our days.

The Rear of the Field

The plants described above are probably the most familiar Cape bulbs
as far as modern European gardeners are concerned. A few others still
trot gamely along in the wake of the gladioli, freesias, and so on, however,

and might be expected to put on a burst of speed as the relentless race for new garden subjects continues. Among those that may become more conspicuous and available before long are the so-called Cape hyacinths, the lachenalias.

According to *The Florists's Guide* of 1825, lachenalias are:

> tender little bulbs, natives of the Cape of Good Hope. There are supposed to be in all, about forty species and varieties. Those most cultivated with us, are the lachenalia quadri-colour, and the tri-coloured, which are very beautiful when in full bloom, exhibiting various colours, on a stem of from six inches to a foot in height, and much in the character of Hyacinths.

The Keeper of the Kew Herbarium, William Hemsley declared that:

> They are easily cultivated, easier perhaps than Hyacinths, and it is not necessary to import a fresh supply of bulbs from Holland every season. Moreover, their flowers appear in winter and early spring, when the less elegant Hyacinths form too prominent a feature in conservatory decoration.

It is certainly rather a mystery as to why lachenalias have not become more popular; there are plenty to choose from, with more than 80 species in the Cape flora, over 20 of which are considered suitable for cultivation indoors in pots, or outdoors in the garden where the climate permits.

One plant that does appear to be making a tentative comeback is the

Kniphofia
Better known as red-hot pokers, and named after German botany professor Johann Hieronymous Kniphof (1704–1763). Of the almost 50 species found in southern Africa, four occur in the Cape. One of these, *K. uvaria*, was the first to be introduced into Europe in 1707. Many hybrids have since been developed which, despite their warm-temperate origins, are frost-hardy and grow comfortably in British gardens.

Pineapple lily *Eucomis*
Four species of the pineapple lily occur in marshes and damp grassland in the eastern part the Cape Floral Kingdom. A striking and trouble-free bulb, it is increasing in popularity among gardeners. This is *E. autumnalis* 'White Dwarf', the species being first grown in England under a previous guise of *E. undulata* at Philip Miller's Chelsea garden from seeds received from the Cape in 1760.

Watsonia
The genus *Watsonia* was named by Miller after Sir William Watson (1715–1787), a physician to the Foundling Hospital, London, and an early promoter of the Linnaean system of naming plants and animals. *Watsonias* have been widely cultivated in Europe and elsewhere since their introduction in the mid-eighteenth century.

belladonna lily *Amaryllis belladonna*. This species first appeared in Europe in the early seventeenth century and was illustrated by Giovanni Battista Ferrari in 1633. In 1753 Linnaeus, charmed by its blushing, feminine prettiness, named it *Amaryllis* after a shepherdess in Ovid's pastoral works (or, indeed, any classical rustic maiden; she also appears in Virgil's bucolic poetry), and *bella donna*, literally 'beautiful lady'.

The name *Amaryllis* is often applied to hippeastrums, those hulking, trumpet-shaped blooms that are so popular at Christmas. This nomenclatural confusion has its origins in the age of Robert Sweet. He gave the provenance of *Amaryllis belladonna* as 'W. Indies' and listed it among many South American species that are now recognised as belonging to the genus *Hippeastrum*. Following much debate, a decision by various clever people in 1987 assigned the name *Amaryllis* exclusively to *A. belladonna* and the only other member of the genus, *A. paradisicola* (a species from the northern Cape that was first encountered in a leafy state in 1972, but not seen flowering until 1997).

Sweet also lists more than 80 'Amaryllis' hybrids that were interspecific *Hippeastrum* combinations, or *Hippeastrum* crossed with other amaryllids such as the cosmopolitan *Crinum* species. The genuine belladonna lily itself has been successfully crossed with *Brunsvigia josephinae*, *Crinum moorei* and *Nerine bowdenii*.

Whatever their parentage or pedigree, the various hybrids do indicate how popular the flowers were in the nineteenth century. Since then, the almost invariable misnomer 'amaryllis' has stuck in the minds of stubborn nurserymen when applied to a range of lilies that are not belladonnas. Also entrenched is the name Jersey lily, because *A. belladonna*, like the nerines on neighbouring Guernsey, has happily embraced life in the wild on this Channel Island and, indeed, in many other temperate parts of the world.

Despite the outstanding success and popularity of the bulbs discussed above, there are numerous equally attractive Cape bulbs, including the majority of the 300 listed in Sweet's *Hortus Britannicus*, that have all but vanished from the horticultural scene since their Georgian and Victorian heyday. Nevertheless, I'm confident that many of these will reappear and a successful transition from veld to flowerbed will, in time, be achieved by a multitude of exciting new introductions from the Cape.

Of course, as residents and visitors over the years have continually affirmed, there is nothing quite like seeing these particular plants in the wild. The South African garden writer and historian Dorothea Fairbridge, for example, wrote in 1924:

> Cape bulbs have a world-wide reputation, and their long Latin names are borne across the earth in florists' catalogues from Haarlem to Philadelphia and back. But none can know their glory fully who has not seen them on the veld, glowing and quivering, under the blaze of a South African sun on a still September mid-day. Spring is in the air. The sundews form a glittering carpet from which grow Ixias and Babianas, Sparaxis, Tritomas

[aloes] – I could go on indefinitely with my list, but a string of names conveys very little of the loveliness which can only be realised by those who have seen it.

For most of us, the nearest we can get to this is probably a bunch of supermarket freesias, but at least it provides a hint of the floral riches and beauty that prevail in the Cape, the 'bulb capital of the world'.

opposite
Lachenalia

These figured in Simon van der Stel's expedition report of 1685 and a number of species were sent back to Europe at that time. A century later the younger Jacquin described the genus from specimens growing in the Schönbrunn glasshouses, naming it after Werner de Lachenal (1739–1800), professor of botany at the University of Basel. Lachenalias are known in the Cape as 'viooltjie' meaning 'little violin' and describing the high-pitched squeak made when the stems are rubbed together.

Top left, *L. unicolor.* A very variable species from the western Cape. Some of its forms are pleasantly scented, and it is easy and rewarding to grow in a cool greenhouse.

Bottom left, *L. aloides* var. *vanzyliae.* One of a number of varieties of *L. aloides,* the most widely cultivated of the lachenalias, this one was introduced to Kirtstenbosch by a Mrs van Zyl in 1927.

Top right, *L. mathewsii.* Named after Joseph Mathews (1871–1949) the first curator of Kirstenbosch. Its renosterveld habitat having been almost completely lost to agriculture, the species is known only from a single location on the Cape west coast. Having not been seen in the wild for over 40 years, it was considered to be extinct until its rediscovery in 1983. Luckily, it thrives in cultivation.

Bottom right, *L. moniliformis.* Another excellent performer in cultivation, but also known only from a single location in the wild.

right
Belladonna lily *Amaryllis belladonna*
first appeared in Europe in the early seventeenth century. The name continues to be sweepingly misapplied by the horticultural trade to *Hippeastrum,* those hefty bulbous perennials from Central and South America.

Five of the almost 700 species of Cape heath. Left to right, *Erica cubica*, *E. blenna*, *E. fairii*, *E. cerinthoides*, and *E. parilis*. Named in 1894 after its discoverer, C. B. Fair, a member of the treasury of the old Cape Parliament, *E. fairii* is extremely rare and confined to a hectare or two of the Cape Peninsula. *E. cerinthoides*, in contrast, occurs widely in southern and eastern South Africa.

CHAPTER TEN

Cross-Pollinating in Upper Tooting

While Cape bulbs have enjoyed mixed fortunes over the years, there are two groups of Cape plants which, having reached dizzy horticultural heights in the years immediately following their introduction, have almost completely vanished from the gardening scene, at least in Britain. They are also the families for which the Cape is best known, one for its extraordinary wealth of species, the other for the bizarre nature of its blooms. They are, respectively, the ericas (heaths) and the proteas.

Plants to Love

Robert Sweet listed 476 ericas in cultivation in Britain in 1826. Of these, 458 were Cape species or cultivars derived from them. Today you'd be lucky to find a single one, cultivar or species, in this country beyond the confines of our national botanical gardens or very occasionally in the greenhouse of a specialist gardener.

At the last count, there were 658 species of *Erica* in the Cape Floral Kingdom. This figure, from the year 2000, would have been out of date almost before the ink was dry as new species are discovered with impressive frequency. No one can predict where the list will end, given that many of the new species look rather like existing ones and others have extremely small ranges in remote and seldom visited areas. The main point is that there is an enormous natural variety to choose from.

Cape heaths were very popular amongst the aristocracy, nurserymen and professional gardeners in the late eighteenth and early nineteenth centuries. Though there is not much evidence that they were adopted by the general gardening public to the extent that, say, freesias or gladioli were. This was because of the difficulty, perceived or real, of growing them.

In 1834, Mr Robertson, 'Nurseryman of Kilkenny', wrote:

Few see ericas in their native perfection: stunted and impoverished, a great proportion of those preserved in our greenhouses must be rather considered as botanical specimens than ornamental plants; and it requires no small amount of skill and attention (both which they unremittingly demand) to keep them alive.

Robertson identified the problem in his part of the world (Ireland) as being not so much the cold as the damp.

Meanwhile, Mr T. Rutger, a regular correspondent with the various gardening publications of the time and clearly a great enthusiast of Cape plants, grew a large number of ericas in his Cornish garden. Commenting on the 'causes and prevention of mildew in Cape heaths' he described how he erected a temporary shelter over his precious plants in winter, removing it in April so that 'during the summer the plants grew rapidly, presenting a most beautiful foliage, with flowers of a very superior character, and, consequently, were much admired by all who saw them.'

Rutger mentions over 20 species, most of them large, colourful, tubular-flowered ones such as *E. cerinthoides*, *E. mammosa*, *E. sparmanii*, and *E. verticillata*. (The latter has since become extinct in the wild at the Cape). He did not think that many would survive outdoors over the winter

---

Throughout the grassy mountains which the hunter must traverse in following this antelope, his eye is often gladdened by romantic dells and sparkling rivulets, whose exhilarating freshness strongly and pleasingly contrasts with the barren, rocky mountain heights and shoulders immediately contiguous. The green banks and little hollows along the margins of these streamlets are adorned with innumerable species of brilliant plants and flowering shrubs in wild profusion. Amongst these, to my eyes, the most dazzling in their beauty were perhaps those lovely heaths for which the Cape is justly renowned. These exquisite plants, singly, or in groups, here adorn the wilderness, with a freedom and luxuriance which, could the English gardener or amateur florist behold, might well feel disheartened, so infinitely does Nature in this favoured clime surpass in wild exuberance the nurselings of his artificial care.

Roualeyn Gordon-Cumming, *Five years of a hunter's life in the far interior of South Africa, with anecdotes of the chase and notices of the native tribes, 1855.*

CROSS-POLLINATING IN UPPER TOOTING • 153

without protection, even in the relatively mild south-west of England.

It's not altogether surprising that, like the majority of Cape plants here, and certainly almost all the perennials, the heaths fared much better indoors than out. If the winter cold didn't get them, then rain and damp at any time of year would. Under glass, however, many people were having more success. For example, in 1826, George Dunbar, Professor of Greek at the University of Edinburgh, could boast no less no less than 344 *Erica* species or cultivars in his 'heath-house' (an integral part of a 30 m long conservatory with 'an abundance of light').

Thirty years later, heaths were still a popular subject with the indoor horticulturist. *The Florist, Fruitist and Garden Miscellany* of 1856 proclaimed 'How beautifully compact and ornamental in their growth are most of our varieties of Cape Heath!' Perhaps surprisingly, given the decline that they were about to experience, the writer also enthuses that:

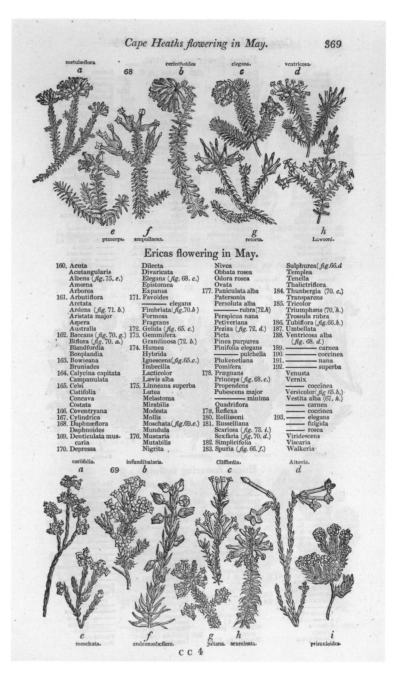

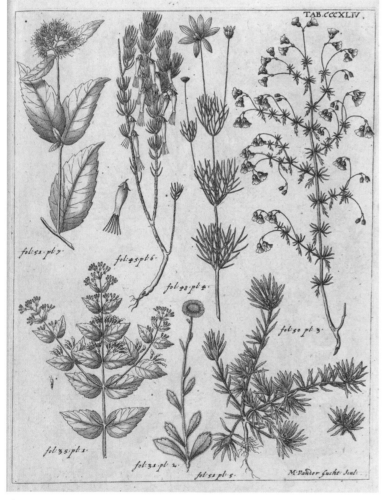

above
Leonard Plukenet's *Almagesti Botanici*, published in 1700, features a number of Cape species, among them *Erica plukenetii* (top, second from left), named for him by Linnaeus in 1753. Plukenet (1642–1706) was physician at Westminster, superintendent of Hampton Court garden, and botanist to Queen Mary.

right
'Art. II' in *The Gardeners Magazine* of 1826 is entitled 'List of Cape Heaths which have been in Flower in Tooting Nursery in each Month of the Year.' and was 'Communicated by Messrs. Rollison, Nurserymen, Tooting.' Described as 'one of the most elegant families of plants', 285 ericas are listed, demonstrating just how popular and profuse they were in cultivation at the time.

They are easily grown, too, and remain a long time in blossom, which, with a few well selected kinds, may be kept up nearly the whole year round. Although indoor gardening loses much of its interest in summer, when Nature is so prodigal of her beauties in the open air, still, even then, pleasure may be derived from an inspection of the section of Cape Heaths which flower at that season; and in winter and spring, when there is little in the way of flowers to induce us to stray beyond the walls of our little greenhouses, these afford us a source of real delight; for who can look upon their deep green leaves and charmingly polished waxy blossoms of various forms, when there is little else to cheer us, without a feeling of satisfaction?

Their fall from grace, if such it was, was precipitous and speedy. In 1897, *The Flower Grower's Guide* commented how 'Ericas are undeniably beautiful when grown … In past times large collections of heaths were grown in many gardens, but are rarely met with now.' The article goes on to say that 'Handsome specimens may be seen at exhibitions, the result of twenty years and more cultural care.' It is interesting to note, however, that at this time 'free-growing winter and spring flowering species' were still 'grown in their thousands in flower markets, and few plants are more chastely attractive during their flowering season.' More in hope than faith, one suspects, the author provides instructions on how to cultivate the heaths, some of which 'will stand a considerable amount of rough treatment, including attempts to repot them by novices.' Of the harder-wooded species he laments that 'the slightest mistake made with these, either in potting or watering, is liable to lead to the death, slow but sure, of the plants.' He lists 24 'easily grown' species or cultivars and 17 exhibition varieties. While a respectable list by today's standards, it is a far cry from the 'ericamania' that had once gripped the nation's fashionable gardeners.

Despite their abundance in the areas frequented by the first seafarers and settlers, Cape heaths took some time to make their way north to Europe. This is simply a reflection of the fact that they were unlikely to survive being uprooted and sent back alive, and the small seeds required specialist interest and techniques to collect.

The first incontrovertible evidence that a Cape *Erica* had reached Europe and was successfully propagated there comes in the form of *E. curvirostris*, a common enough species in the area of the Dutch settlement, depicted by Maria Moninckx sometime between 1686 and 1706. Contemporary appearances of ericas in print include Plukenet's *Almagesti Botanici Mantissa* of 1700, in which three species are depicted. These illustrations were almost certainly based on dried rather than living specimens.

Progress was still slow, and it wasn't until the 1730s that these southern heaths had successfully established a root-hold in Europe when Linnaeus recorded five species growing in George Clifford's Hartecamp garden. Thereafter, increasing numbers were added to the list and by the end of the eighteenth century almost 150 Cape species had been scientifically described, a major contribution being made by Thunberg who named about 40.

Although new species were appearing with what must have been, to the European botanists used to only a paltry handful of heaths in their own part of the world, startling frequency, there seemed to be no great urge to cultivate them. This situation changed radically, however, with Masson's arrival at the Cape. Between 1787 and 1795 Kew received the seeds of about 86 species of *Erica* from Masson and the garden's total in cultivation rose to about 140, there having been 41 in 1789 and 80 in 1792.

By the nineteenth century, the *Erica* situation was looking decidedly brighter, metaphorically and literally – improved glasshouses, well lit and well ventilated, suited them in just the way that dingy, dank environments hadn't. And, for whatever reason, the genus had captured the imagination of the private horticulturists and their commercial suppliers so that between 1798 and 1802 some 275 new species or cultivars were described.

Further supplies were provided for the aristocracy and the nurserymen by Paterson, Burchell, Scholl and Boos, the latter two being noteworthy because they successfully sent live plants back to the royal garden in Vienna. The pair don't seem to have introduced any new species into cultivation, however, their horticultural thunder having been stolen by the assiduous Masson.

James Niven, who had spent many years at the Cape (see Chapter 7) fared rather better, introducing 12 new species to cultivation through his consignments to George Hibbert, the Empress Joséphine, and Lee and Kennedy's nursery. Inevitably, however, and despite the richness of the Cape heath flora, the law of diminishing returns eventually kicked in, the number of new species arriving in Europe declined, and by 1810 had virtually fizzled away to nothing.

The passion for Cape heaths did not, however, die out there and then; with plenty of material to work on, nurserymen shifted their attention to hybridisation by playing the role of pollinating sunbird or insect.

One of the first to take up this challenge was William Rollisson at his Springfield Nursery in Upper Tooting, Surrey, although for reasons of commercial confidence he initially kept tight-lipped about his work. By the time of his death in 1842 he had created at least 90 hybrids including, according to his friend W. H. Story, 'most (I was going to say *all*) of the choicest and most favoured varieties now in cultivation.'

In 1826 two lists of Rollisson's heaths were published in *The Gardener's Magazine*. These presumably represent Rollisson finally 'going public', and an impressive debut it is, with 285 'species and varieties described and presented according to the months in which they flower'.

Interestingly, the lists follow an article by James Bowie entitled 'Hints for the better Cultivation of the Cape Heaths, derived from Observations of their Nature, Soils, and Situation.' This was one of Bowie's very few contributions to the literature and preceded his generally calamitous and unproductive return to the Cape following his earlier success there as a collector. Noting that he was shortly to return, the editor's hopes that the magazine would be 'favoured with occasional communications from him on heaths, Proteas, and other subjects connected with that interesting colony' were, sadly, unfulfilled.

While the commercially canny Rollisson was busy cross-pollinating in Upper Tooting, Dean Herbert had been pursuing this interest as well. His first success was a cross of *E. vestita 'coccinea'* and *E. jasminiflora*. The cultivars that he obtained he described as 'most distinct and remarkable, the individuals of each species being perfectly uniform.' Unfortunately, all his ericas seem to have died soon after his move to Yorkshire in 1814.

Herbert was an astute and inquisitive observer, as already demonstrated by his work with Cape bulbs. He noted how the pollen in the tubular-flowered ericas could only be released by poking them with a pin in the way that a sunbird or a long-tongued insect would do in the wild. He concluded that the genus 'affords greater facility of intermixing' and went on to prove his point by crossing two species that another eminent botanist, Richard Salisbury, had considered belonged to separate genera and were, therefore, reproductively incompatible.

This ease of crossing was one of the attributes that made Cape heaths so popular. By 1842 it was observed that the garden hybrids 'are now multiplied almost indefinitely, so as to render fruitless any attempt at describing them on paper.' The situation in the late eighteenth and first half of the nineteenth centuries is further confirmed by botanical historians Charles Nelson and Ted Oliver as 'a vast assemblage of wild-collected, seed-raised species, multifarious hybrids and, undoubtedly, innumerable backcrosses.'

This proliferation and dodgy nomenclature were no discouragement to the gardeners, however, and the frequency of their appearance in the literature reflected the passion with which heaths were grown, at least by aristocrats and nurserymen.

Prominent among the professional enthusiasts was William McNab who, as superintendent of the Royal Botanic Garden in Edinburgh, had the luxury of superior resources and facilities at his disposal. In 1832 he wrote *A treatise on the propagation, cultivation and general treatment of Cape heaths*. He wasted no time in putting the Cape heaths into a local context:

> As Scotland is a country already famed for its native heather, it may appear somewhat paradoxical for anyone to attempt to recommend methods for the cultivation of any sort, where much expense and labour is annually expended to get a part of the genus eradicated.

It is also paradoxical that while Scotland, having failed to eradicate 'a part of the genus', now makes a great song and a dance (literally and metaphorically) over its heaths but can boast only a tiny sprinkling of species. Furthermore, its nominally bonny purple moorlands are artificial environments that are almost entirely the consequence of forest destruction on a monumental scale, chronic overgrazing by sheep and deer, and land management that promotes a solitary member of the family Ericaceae, being heather *Calluna vulgaris*, for sustaining the red grouse-shooting industry.

McNab records how some of the finest erica collections had deteriorated or disappeared over recent years 'From the supposed difficulty in management', a situation he greatly regretted and strove to overcome.

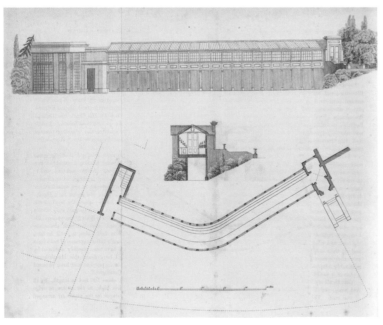

above left
William McNab (1780–1848), an authority on Cape erica cultivation, worked at Kew before being made superintendent of the Edinburgh Royal Botanic Gardens in 1810. His appointment was approved by Banks after wrangling over his annual salary which was finally set at 'some shillings below £50 just to avoid the Tax'. A 'strange and interesting [Cape heath] ... so different from all others' was placed in its own genus *Mcnabia*, then reassigned to *Erica* with the specific name *nabea* retaining a semblance of the man it honoured. It could 'be easily grown in a relatively damp situation as a botanical oddity.'

left
The Duke of Bedford's Cape heaths also featured prominently in a catalogue of the plants growing in the gardens. Written by the Duke's gardener, James Forbes (1773–1861) and published in 1833, *Hortus Woburnensis* listed more than 400 Cape ericas and included an engraving of the heathery. The catalogue also reveals that the Woburn greenhouses and gardens contained almost 80 species of Cape bulb, 20 proteas, 3 strelitzias, 75 mesembryanthemums, and 30 species and 350 cultivars of pelargoniums.

One of the finest collections of Cape heaths ever assembled was at the Woburn estate of John Russell, Sixth Duke of Bedford (1766–1839). An account of these, *Hortus ericæus Woburnensis: or, a catalogue of heaths in the collection of the Duke of Bedford at Woburn Abbey*, was prepared by the Duke's gardener George Sinclair (1787–1834) and published in 1825. Born on the Scottish Borders estate of Mellerstain, where his father was gardener, Sinclair joined the horticultural staff at Woburn in about 1807, before setting up his own nursery business at New Cross, Surrey in 1825.

Despite McNab's concern about the plants' decline, they were still very popular when he was writing, at least in certain sectors. Some of the wealthiest landowners constructed glasshouses specifically for Cape heaths, the Duke of Bedford's at Woburn, for example, being 'above 11 feet in length, by 12 feet wide, and nine feet high.' The *Hortus Woburnensis* of 1833 describes in detail the housing and cultivation of these plants. A window of the heath-house was decorated with 'about 50 of the most beautiful flowering species … executed by Mr. Andrews, and so accurately done, that they can scarcely be distinguished from living plants.'

'Mr. Andrews' was Henry Andrews, who in 1802–09 produced a series of four volumes entitled *Coloured Engravings of Heaths, the drawings taken from living plants only, with the appropriate specific character, full description, native place of growth, and time and flowering of each.* The volumes contain 288 hand-coloured engravings and were followed, in 1804–12, by the five-volume *The Heathery; or a Monograph of the Genus Erica*, containing 300 plates.

Cape heaths were also illustrated at Kew by Franz Bauer, 'Botanick painter to His Majesty' in William Aiton's publications and, together with Andrews's works, these superb illustrations must have contributed to the growing interest in these particular plants and their subsequent popularity over the next half century. In 1846 a correspondent to *The Floricultural Cabinet* solemnly declared that a 'heathery, to be complete, should contain 250–300 species'!

By the late 1800s, however, Cape heaths had all but vanished from the scene, victims of fashion and modern technology that made greenhouses more like Turkish baths and less the warm, dry, airy places that ericas and their Cape compatriots required. Given that many hundreds of hybrids had been developed, including a 'gumless' one, '*E. x dennisoniana*', that had eliminated the stickiness characteristic of many of the attractive tubular-flowered species, this was regrettable. Only a fairly healthy pot-plant market briefly persisted, the Christmas sales accounting for some 200,000 potted heaths 'of easily propagated varieties' in the late 1870s. This trade lasted into the early years of the twentieth century before following the private collections into an all-but terminal decline.

Cape heaths are very liable to be attacked by mildew, particularly in the neighbourhood of London; and some collections have been nearly destroyed from this cause. Sulphur, applied either in a dry or moist state, is the most effectual cure, and should be applied upon the very first appearance of the disease, by dusting the plants all over with the dry flour of sulphur, or by making up a thick lather of sulphur, mixed with soap, and laid on the plants with a painter's brush.

Heaths are not very subject to the attacks of insects; the green fly, however, sometimes assails them, but these are readily got rid of by slight fumigations of tobacco.

Charles McIntosh, *The Greenhouse, Hot House and Stove*, 1838.

The demise of the heaths was lamented by Joseph Hooker in 1874:

> Many years ago, the Cape heaths formed a conspicuous feature in the greenhouses of our grandfathers, and in the illustrated horticultural works of the day … These have given place to the culture of soft-wooded plants – Geraniums, Calceolarias, Fuchsias, &co.; and the best collections of the present day are mere ghosts of the once glorious ericeta of Woburn, Edinburgh, Glasgow, and Kew. A vast number of the species have indeed fallen out of cultivation … No less than 186 of Erica were cultivated [at Kew] in 1811, now we have not above 50, together with many hybrids and varieties.

Perhaps there is still hope for the ericas. In *The Florist, Fruitist and Garden Miscellany* of 1856, a writer muses that there is:

> one point to which little attention has apparently been paid, and that is, the crossing of our tender heaths with hardy ones. It strikes us, and the idea is not new, that there should be no impracticability in crossing the splendid species of South Africa with the hardy natives that adorn our shrubberies. We need not say that a successful result in that direction would be most interesting, and the additional beauty that would be thus introduced to our beds and borders would more than compensate for the trouble which such an attempt would cause. That much may be done by hybridisation has of late years become marvellously manifest; and surely there can be no obstacle to a union between the tender sorts and the little hardy varieties which we find everywhere in flower. Let the experiment be but fairly tried by skilful hands, and we have little fear of the result.

Who could resist this challenge on being reminded of the praise lavished on the great tribe of Cape heaths by the 'able conductor of the *Gardener's Magazine*'?:

> Of what other genus can it be said, that every species, without exception, is beautiful throughout the year, and of every period of its growth, in flower, or out of flower, and of every size and age? Suppose an individual had the penance imposed on him, of being forbidden to cultivate more than one genus of ornamental plants, is there a genus he could make a choice of at all to be compared to Ericaeae, perpetually green, perpetually in flower, of all colours, of all sizes, and of many shapes?

Variable and Different

In the Aegean deep there dwells a seer
The sea-blue Proteus…
… when you hold him fettered in your hands,
The changing forms of beasts will mock your grasp.
He'll suddenly become a bristly boar,
Or deadly tiger or a scaly snake,
Or change into a lion with a tawny mane,
Or blaze up into fire and thus escape,
Or melt into thin water and be gone.
The more he changes shape, the tighter you,

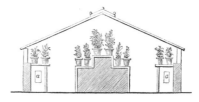

The heath family constitute an extensive assemblage of low shrubby evergreen plants, much valued for the beauty of their flowers, and the blossoming of them in the winter season. Scarcely any exotic heaths were known in Miller's time, and none of the Cape species. Almost the whole of these have been introduced to Europe during the reign of Geo. III, and the greater part by Masson, a collector who made two voyages to Africa at that king's expense.
The colours of their blossoms are so various, and their shapes so endless and so chaste. What a combination of classic forms might not an ingenious artist invent from a study of their beautiful tubular and vase-like corollas! The colour and character of their foliage must attract and interest. In fact, they realise the description which a lady in my hearing once gave them, they are 'plants to love'.

J. C. Loudon, 1835.

right
'Propagating ericas' from John Wright's *The Flower Grower's Guide*, published in 1897. *RHS, Lindley Library*

below
A 'Heathery' from *The Greenhouse, Hot House and Stove*, by John McIntosh, 1838.

below right
Erica pot from *The Greenhouse, Hot House and Stove*.

Most surprising is it in this age of botanical enterprise, with the high estimation in which gardeners hold Cape Heaths for decorating the conservatory and greenhouse all the year round, to find that no attempt has hitherto been made to cross them and their numerous garden varieties with the hardy European kind, so as to produce a perfectly hardy race suited to the open border. If the art were difficult, and the chances hopelessly rare, this might readily be accounted for; but as there are ample materials, and great facility, no fear need be entertained that a person with ordinary perseverance, and a slight botanical acquaintance with plants would fail.

George Gordon, *The Journal of Horticulture, Cottage Gardener, and Country Gentleman*, 1863.

right
Members of the protea family from Johann Weinmann's *Phytanthoza icongraphia* (1737–45). Plate 899, tree pincushion *Leucospermum conocarpodendron* and *Mimetes argenteus*; Plate 905, a conebush *Leucadendron* and *M. cucullatus*.

Masson noted how 'The [tree pincushion] grows on the skirts of the mountains … The people of Cape Town use it for burn wood which is dug up and carried by slaves for 5 or 6 miles on a stick about 5 feet long, with a large bundle on each end, which they carry over their shoulders. It would make a fine ornament among the green house plants in Europe …'

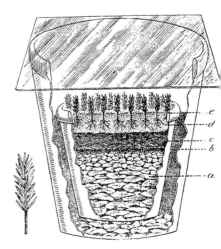

*a. Scolymocephalus Africanus argenteus foliis Dory-
cam. B. b. Scolymocephalus Africanus folio crasso
nervoso, Holo-braun. B.*

*a. Scolymocephalus   seu Conocarpodendron folio tenui angusto. B.
b. Scolymocephalus seu Hy-   pophyllocarpodendron polis tribus in summitate. B.*

My son, must hold the bonds until at last
He once more takes the shape you found him in.

Virgil, *The Georgics*, Book IV (*c.* 30 BC.)

Amongst the plethora of ancient Greek divinities was this one, Proteus, whose official occupation was shepherding marine beasts for the sea god Poseidon, but who also had the gift of prophecy as a useful sideline. As Virgil describes, he could change his shape at will, which is probably why Linnaeus chose this name for what is arguably the Cape Floral Kingdom's quintessential and best known plant family, as it, like Proteus, occurs in diverse shapes and forms. So, in describing 'proteas' it must be remembered that *Protea* is just one genus in the family Proteaceae which, in the Cape Floral Kingdom alone, comprises some 330 species belonging to 14 genera, including plants such as the conebushes *Leucadendron* and pincushions *Leucospermum* in addition to the more familiar proteas or sugarbushes.

Sweet listed almost 150 protea species in cultivation in the early nineteenth century, but not all of these were necessarily widely grown. There was no doubt, however, that proteas in their multitudinous shapes and forms had caught the imagination of those who had the wherewithal to obtain them and the facilities in which to grow them. Like the Cape heaths, and often sharing their warm, well-ventilated quarters, they were the height of fashion as well as being arresting botanical curiosities. Their ascendancy to the heights of botanical chic did, however, take some time.

Following the arrival of the 'thistle from Madagascar' in 1597, it was almost 100 years before the second appearance of a protea in Europe. This took the form of six species from Table Mountain collected by Paul Hermann in 1672.

One of the commonest proteas around Cape Town is, or at least was, *Protea repens*, so it was not surprising that this was one of the first to find its way to England. A specimen collected by a Mr Goddard was exhibited, along with bits of a silver tree *Leucadendron argenteum* from the slopes of Table Mountain, at the Royal Society and featured in its *Philosophical Transactions* in 1693. Protea authority and botanical historian John Rourke observes how 'Such tantalising fragments of a bizarre and utterly unknown flora must have been as exciting to learned men in Europe … as lumps of moon rock are to the scientists of today.'

A later caller at the Cape, James Cunningham, made a small collection of proteas when he stopped off on his return from the Far East where he served as a surgeon with the East India Company. His plants included a king protea *P. cynaroides* that featured in Plukenet's *Amaltheum Botanicum* of 1705. The rendition hardly does justice to the colossal, bizarre flowerhead, but it is at least identifiable.

Illustrations of proteas made at the Cape also began to find their way north at this time. A significant set of watercolours, most likely painted by Hendrik Oldenland, superintendent of the Company's Garden, came into the hands of Hermann Boerhaave. He had 24 of these made into copper engravings and included them in his catalogue of the Leiden Botanic Garden in 1720.

Just a few weeks after arriving in Holland in 1735, Linnaeaus published his *Systema Naturae*, including within it Boerhaave's obscure African plants under the name *Protea*. He later justified this name, declaring them to be 'Greatly variable and different like Proteus himself'. While this is undeniably true, and fits the family perfectly, it seems more good fortune than astute judgement that the name Linnaeus gave to the family on the basis of a few illustrations later turned out to be more appropriate than he could ever have imagined. He would certainly have been relieved to learn that the family he christened is, indeed, extremely 'variable and different' and comprises about 1,400 species in more than 60 genera, widely distributed through South and Central America, tropical Africa, Madagascar, New Guinea, New Zealand and Australia. The last-named country can boast 800 species in 45 genera, the most famous of which are *Banksia* and *Macadamia* (of 'nut' fame). The numerous Cape species range from straggly offerings that can barely raise themselves a leaf's width off the ground to handsome trees 30 m high.

This broad natural variety was obviously one of the attractions to the early horticulturists who grew proteas, but it has to be said that the blooms of *Protea* proper must have fair taken the breath away of those who first saw them in flower.

For this they have, in the first place, Francis Masson to thank. He sent *Protea* seeds back to Kew into the methodical care of William Aiton. The first species to bloom was *P. repens*, raised from seed that Masson had collected in 1774. Progress was slow but sure and by 1789 Kew, now under the supervision of William Aiton Jnr, could boast six

right
King protea *Protea cynaroides* is one of the most widespread members of its family and occurs in a range of habitats from the coastal plains to 1,500 m in the high peaks, such as here in the Outeniqua Mountains. *Richard Cowling*

below
At the other end of the scale from *Protea cynaroides* is *P. pudens*, whose inflorescences are only 5–8 cm across. Described by the protea authority, John Rourke, as 'this beguiling protea', it was discovered in 1895. '*Pudens*' means 'modest' or 'bashful', the flowerheads being borne at ground level, tucked within the foliage.

A wealthy merchant and obsessive plant collector, George Hibbert (1757–1837) obtained the rarest and most expensive species and built up an unrivalled range of cultivated proteas at his Clapham home. *Photographic Survey, The Courtauld Institute of Art, London. Private collection*

Although they do not have particularly attractive inflorescences, at least when compared to other members of the protea family, conebushes *Leucadendron* often have bright red or yellow foliage or unusual seedheads. This is *L. platyspermum*, the species name meaning 'flat seed'. The seeds are held in the hard, long-lived cones until the plant dies of old age or, more likely, fire sweeps through the fynbos. The winged seed (at up to 32 mm, the largest in the genus) then germinates within the cone, its expanding root pushing it to freedom. The cones are much sought-after by flower-arrangers for their unusual looks and enduring qualities.

species of *Protea* in flower and, by 1810, a further 16, with two others that could not be induced to bloom. Given their unusual and spectacular inflorescences, it is perhaps surprising that the artist in residence, Franz Bauer, should have illustrated only two species (*P. lepidocarpodendron* and *P. compacta*), and finished neither painting.

The interest generated by the Kew proteas, and the growing obsession with Cape plants in general that seemed to be a feature of the reign of George III, soon spread beyond regal confines. Supplied by the Cape collectors, some of the nurserymen and private businessmen built up impressive collections.

The finest commercial stock was established at the Vineyard Nursery, which became the major supplier of protea plants to anyone who could afford them. Prominent among these was George Hibbert, a wealthy merchant, Member of Parliament, and an Alderman of London who, in his day, could boast a collection of proteas in his own botanic garden in Clapham that was greater even than that of Kew.

In 1798 Hibbert sent James Niven to the Cape to collect on his behalf. Such was the success of Niven's ventures and the skill of Hibbert's gardener, Joseph Knight, as a plantsman that by 1805 there were 35 *Protea* species flowering in Hibbert's glasshouses, together with many other members of the Proteaceae family including *Leucospermum*, *Leucadendron* and *Serruria*. This was a remarkable achievement and such a collection was not equalled or surpassed for another 150 years, and then only at Kirstenbosch National Botanical Garden in the proteas' home patch.

For the next 30 years or so, those who could afford the heating bills were able to enjoy the splendours of this wholly unusual family. Certainly Hibbert and men such as the Duke of Bedford and the Marquis of Blandford were renowned for both their proteas and their heaths.

Interest in proteas was also quite healthy on the continent. Here, their history does not go back as far as Britain as, for once, the Dutch were lagging a little behind their cross-Channel competitors. It was not until 1794 that a recognisable species finally took root in Holland, in the form of *P. aurea* that flowered in the botanic gardens of Utrecht.

In France, Empress Joséphine built up a fine collection of proteas at Malmaison, the product of James Niven's industry, while the rest of Europe was trying to resist the advances of her husband. The end of the Napoleonic Wars finally allowed the Cape-plant aficionados of Europe to make up for lost time, and a number of collectors headed for South Africa to satisfy this resurgence of interest. The efforts of German collectors, for example, allowed protea cultivation to reach unprecedented heights in that country. By 1821 the Berlin Royal Botanic Garden had 18 species in cultivation, and various minor royal figures could boast glasshouses enhanced by these Cape beauties. In Russia proteas were also to be found, suitably cosseted, in the Tsar's Imperial Garden at St Petersburg.

Despite these successes, protea cultivation in mainland Europe pretty much followed that of Britain into oblivion in the 1830s or thereabouts.

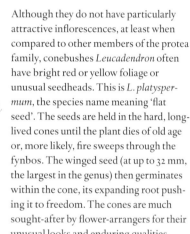

Changing tastes, and steam-heating, whose high temperatures and humidity ultimately killed off the Cape heaths in cultivation, also put paid to the proteas in an even shorter time. Echoing his sentiments regarding the Cape heaths a few years earlier, Joseph Hooker penned a belated obituary to the proteas in 1881:

> The Cape Proteaceae, the favourites of our grandfathers, may be said to have 'gone out of cultivation', so completely have they been replaced by other tribes … this is mainly due to the introduction of those improved systems of heating houses and that incessant watering, that favours soft-wooded tropical plants and is the death of the Proteas of South Africa and the Banksias of Australia. Nevertheless, that these and many others requiring like treatment will be reintroduced, and will be the wonders of the shows of many successive season, is as certain as they that once were the glories of the old hot-air heated kilns that our forefathers called stoves, in which Orchids quickly perished and Banksias and Proteas throve magnificently.

Hooker has yet to be vindicated, at least in one respect – there are very few proteas in cultivation under glass in Britain today. He was, however, right about the return of the 'wonders' that are the proteas, but not in the circumstances that he had predicted.

Familiarity, it is true, breeds contempt, and while the Hibberts of the north had been lavishing time and money on cultivating precious proteas, the plants were almost entirely ignored back in the Cape. Well, at least as far as aesthetics and botanical interest were concerned, but where any practical use could be made of them it was. Wild almond *Brabejum stellatifolium* was planted for barriers; the wood of waboom ('wagon tree') *Protea nitida* was used for a variety of purposes ranging from wagon-wheel rims to tobacco pipes, and its bark stripped to make tannin for tanning leather and brewed up as an astringent; nectar was drained from *P. repens* inflorescences to make a sugary syrup ('bossie stroop'), and any bush that was even remotely big enough was hacked down to make implements, furniture or building material or burnt for cooking and heating. It's not that any of these were particularly good for the purpose for which they were exploited, it's just that there wasn't much else to fall back on.

By the twentieth century, however, an interest in proteas as things of fascination and beauty, as opposed to footstools and firewood, had arisen in the Cape. Certainly there was a blank canvas as far as cultivation here was concerned, one horticulturist claiming in 1919 that he had never seen one in a garden. He was not alone in being bemused by this, partly due to their popularity in Europe, and partly the fact that many proteas are, quite simply, spectacular. The reason, upon which Dorothea Fairbridge speculated, was that there was not much inclination to grow something that flourished like a weed just over the fence. Another was the predilection for the classical English garden look, with roses and hollyhocks and acres of carefully manicured lawn, a style ill-suited to the environmental conditions of the Cape.

The modern age of protea appreciation in the Cape began in the spring of 1920 when a display of proteas at the Visitors' Bureau in the centre of Cape Town attracted thousands of admirers. These proteas had been raised at the youthful Kirstenbosch, a scheme encouraged by its enterprising curator, Joseph Matthews. The following year, he published a short article entitled 'The cultivation of proteas and their allies' in the journal of the Botanical Society of South Africa and, not long afterwards, protea seeds were offered by Kirstenbosch to members of the Society.

With proteas now elevated above the rank of weeds, horticulturists began to respond to the demand for seeds and plants, albeit it on a small scale to begin with. A major development occurred when harvesting from the wild was recognised as undesirable, not to say unsustainable, and was declared illegal under Official Ordinance in 1937. This had, by all accounts, long been an uncontrolled and random occupation that had seen the populations of many plants, but mainly proteas and ericas, severely depleted. From the conservation point of view, therefore, growing your own and selling the flowers was a more acceptable way of satisfying the demand for cut stems. This reflected an awakening interest in nature conservation generally in South Africa, it having been seen as unnecessary in a country whose natural resources had, for the past three centuries or so, been considered limitless, uninteresting, or both, by the majority of settlers.

The first commercial cultivator of proteas at the Cape was the redoubtable Miss Kate Stanford who established a nursery near Stellenbosch in the 1920s and '30s. Some years later, Frank Batchelor, a fruit and flower grower, began to diversify into protea cultivation on his property, also not far from Stellenbosch. He not only provided seeds and plants for the garden, but was able to extend his menu of cut flowers beyond traditional favourites like gladioli. At his farm, subsequently named 'Protea Heights', he successfully harvested a crop of protea stems in 1949 and, using the relatively recent innovation of air-freight to best advantage, exhibited the choicest blooms to great acclaim at the spring show of the Transvaal Horticultural Society in Johannesburg. With the upsurge in enthusiasm for the plants as garden and cut-flowers, improved methods of cultivation were adopted and hybrids were developed.

opposite
Specimens of sugarbush *Protea repens* and silver tree *Leucadendron argenteum* were exhibited at a meeting of the Royal Society, and reported by Hans Sloane in its Philosophical Transactions of 1693: 'Alderman Charles Chamberlain having favoured the Royal Society to present them (among other natural rarities) with two Branches of Trees and their Fruits, brought from the Cape of Good-hope by Mr. Goddard, which being very curious, and are not yet any where perfectly described, it was thought fit to publish their figures and Descriptions; … the first of them has been brought hither for its Beauty in Pots, as well as raised in England from the seed brought from the Cape of Good-hope, where it is called the Silver Pine … (and) I am assured by Mr. James Pettiver, that it is planted by the Dutch in their Famous garden, being thought one of its greatest Ornaments.'
The Royal Society

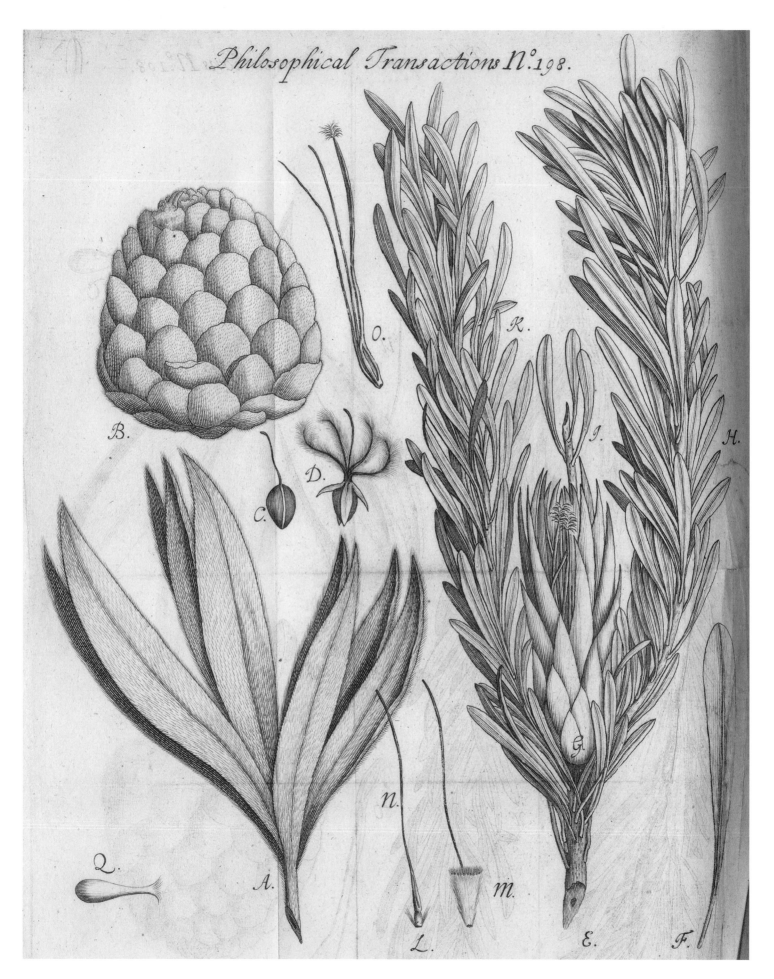

With their unworldly appearance, proteas look sufficiently extraterrestrial to feature in fantasy and science fiction, including Star Trek™. In 'Offspring', a poignant and philosophically challenging episode from *Star Trek: The Next Generation*, the android Data created Lal, his ultimately doomed 'daughter', aboard the U.S.S. *Enterprise*™. Here they admire a *Protea speciosa* with a *P. roupelliae* just appearing in the centre foreground.

CBS Studios Inc. TM & © 2011 CBS Studios Inc. STAR TREK and related marks are trademarks of CBS Studios Inc. All Rights Reserved. Image search by Marian Cordry, Co-ordinator, Materials Archive, CBS

There are approximately 360 species of South African Proteaceae and at least 70 species are marketed locally as fresh cut-flowers. Most species also make outstanding garden subjects. There is a choice of tall elegant shrubs for height and background. There are procumbent, sprawling prostrate proteas for ground cover or for growing in rockeries. In between these extremes are a wealth of medium-sized shrubs that produce handsome flowers at different times of the year. Other species have colourful foliage with seasonal colour changes ... In more severe climates they may be grown successfully in containers under greenhouse conditions which require the simulation of a Mediterranean-type climate ... Humidity must be kept low and there should be an adequate circulation of cool air. Hot and steamy conditions should be avoided at all costs.

Neville Brown *et al, Grow Proteas*, 2002.

Bought on the concourse of Glasgow Central Station, about as far removed geographically, ecologically and culturally from its natural habitat as it is possible to be, this pincushion still retains a strong resemblance to the original *Leucospermum cordifolium* from which a multitude of cultivars has been developed. These have also featured on Star Trek™ under the guise of 'Antarian Moon Blossoms'!

In 1959 *Proteas: know them and grow them* by Marie Vogts was published, the first book to describe the cultivation of the group since 1809, when Joseph Knight included a section on the subject in his new classification of the Proteaceae. Dr Vogts went further than just cataloguing the wild plants; she hunted down naturally-occurring variants and recorded flowering times to minimise the gaps between blooms, and assessed the cultivation potential and requirements of a wide range of species. She was also instrumental in establishing the Protea Research Unit at the Department of Agriculture and, in the course of 40 years work, laid the solid foundations of the protea-growing industry.

Meanwhile, in the 1950s, Capetonian Ruth Middleman also took up the cause with relish and within a few years could offer 48 species for sale. With her efforts and those of Frank Batchelor and Marie Vogts, proteas had at last arrived in South Africa. This was ironic, because they had never actually left.

Today, protea cultivation is a major industry in the western Cape. An even bigger protea industry, however, is to be found overseas. While South Africans were studiously ignoring the goldmine on their doorstep, their Antipodean and Californian counterparts were busy planting proteas in their gardens as early as the 1890s. Fifty years later, commercial plantations and selective breeding to supply the cut-flower market were under way in New Zealand and Australia, followed by countries as distant and diverse as Israel and Norfolk Island.

In 1964 Sam McFadden, a professor of horticulture, imported protea seeds to Honolulu and within five years had established Hawai'i as a major centre of protea research and cultivation. One of the islands' trump cards was that the pincushion-proteas *Leucospermum* flowered there two months earlier than in California, giving them a highly competitive edge in what was becoming a keenly-contested North American market.

In 1974 the South African Wildflower Growers Association changed its name to the South African Protea Producers and Exporters (SAPPEX) to reflect the mainstay of its market. Its vision is:

> to ensure that an economically stable and financially sound commercial Protea and Fynbos industry exists in harmony with the environment and that the industry contributes on a national basis towards balanced economic growth and prosperity of the broader community and is recognized on the overseas market as an important role player in the supply of quality products.

If adhered to, this is about as good as it gets from the environmental, commercial and social standpoints.

The 1970s also saw the formation of the International Protea Association to facilitate protea-growing on a global scale and to encourage debate and good practice amongst its members. The International Protea Register is maintained by the South African Department of Agriculture to keep tabs on new cultivars and their commercial rights. Certainly the number of cultivars has increased by leaps and bounds in recent years as demonstrated by the University of Hawai'i's Protea Research Project release of no fewer than 102 new pincushion cultivars between 1999 and 2004.

Not only are natural variants being identified and assessed, but crossing species is becoming increasingly popular. Although the pure species look impressive enough on their own, this hasn't discouraged breeders from crossing king protea *P. cynaroides* or queen protea *P. magnifica* with a number of their subjects, which I suppose is a trendy sort of thing to do in this era of narrowing the social gap between monarchy and commoners.

The result of all this is that we can now enjoy proteas from the Cape or their adopted homes at any time of year and in increasing variety. At the peak of the flowering season around 200 tonnes of 'Protea and Cape Greens' are exported from South Africa every week, the bulk destined for Central Europe but with a respectable proportion coming to Britain. 'Cape Bouquets' consisting of proteas and pincushions padded out with fillers like *Leucadendron*, everlasting daisies, and sundry Cape shrubs (some of them dyed lurid shades of blue, pink or green), are now also available in most florists and supermarkets.

Having established themselves in the international cut-flower market, can we now expect proteas to make their way back into British gardens and greenhouses again? If so, then one of the front-runners is likely to be *Leucadendron* x 'Safari Sunset', a conebush developed in New Zealand many years ago and known to be frost hardy. Various attractive and unusual pincushions and serrurias are already ensconced in large pots on patios in warmer parts of the world, so it is perhaps just a question of time before cultivars are developed that can bring a little quintessential Cape character to colder climes.

Proteaceae on display. Cultivated conebushes, proteas and pincushions at a Cape Town flower show.

CHAPTER ELEVEN

This Good-Tempered Flower

Ericas and proteas may have come and gone from cultivation, but pelargoniums, it seems, are here for ever. It is not difficult to understand why they have made the greatest contribution of any Cape plant to worldwide horticulture: they look good, come in a multitude of sizes and colours, are easy to hybridise and multiply, flower profusely and dependably, are relatively undemanding, and will happily grow on any sunny windowsill.

There are about 250 pelargonium species in the world, 148 of which occur in the Cape Floral Kingdom (79 being endemic). Given this diversity, it is perhaps surprising that only about 20 Cape species have given rise to the thousands of varieties developed over the years. Indeed, the most important contribution to pelargonium hybridisation has probably been made by less than half this number.

Modern pelargoniums are classed according to their flower and foliage characteristics, these being a consequence of their horticultural history and, ultimately, the wild species from which they are descended. The general consensus is that the main groups are: species, regal, angel, ivy-leaved, scented-leaved, unique, and zonal, with a group known as decorative added by some authorities. Various subsections have also

been created to accommodate particular variations, some sections have dwarf forms, and zonals can be subdivided into cactus-flowered, double- and semi-double-flowered, fancy-leaved, Formosum hybrids, rosebud, single-flowered, and stellar.

Whatever their nomenclature and status, the cultivated pelargoniums have their roots in the warm, thrifty soils of the Cape, and their transformation from wild flowers into denizens of the greenhouse and conservatory follows a path that has wended its way through almost 400 years of green-fingered fiddling.

The Sad Geranium

Pelargonium triste, a common species on the sandy flats around Cape Town, was the first pelargonium to be introduced to, and cultivated in, Europe, probably because its fleshy root allowed it to be easily collected, stored and shipped. It also sometimes has ferny leaves a bit like those of a carrot and its hefty tuber is beetroot-red inside, so perhaps a few were pulled up and taken on board in the hope that they could be eaten. In fact, the plants are marginally palatable, but more useful medicinally as a root decoction used in the treatment of diarrhoea. If this snippet of indigenous herbal wisdom was made known to the early Cape visitors, it would have been useful to have a supply on board the dysentery-ridden vessels returning from the Far East.

P. triste appeared on the horticultural scene in the early seventeenth century in the Parisian garden of René Morin, from whom John Tradescant obtained seed that gave rise to plants that flowered in his Lambeth garden in 1632. There is some evidence that this species enjoyed a flurry of popularity thereafter appearing, for example, under the name 'Geranium noctu' in the 1677 catalogue of nurseryman William Lucas of Strand Lane, London. Once the novelty had worn off, however, and despite its enigmatic charm, *P. triste* all but vanished as more showy species and, in due course, hybrids appeared on the scene.

---

opposite

'Geraniums', an oil-on-canvas painted in Boston in 1888–89 by the prolific American impressionist Childe Hassem (1859–1935). Cape pelargoniums reached North America not long after arriving in Europe. Their rapid spread cross-country to California found them in an environment very much like the Cape, so they flourished without the need for greenhouse or conservatory.

The Hyde Collection, Glens Falls, New York, 1971.22

Geraniums are to me a sort of test flower, for long experience has told me that people who do not like geraniums have something morally unsound about them. Sooner or later you will find them out, you will discover that they drink or steal or speak sharply to their cat. Never trust a man or woman who is not passionately devoted to geraniums.

Beverley Nichols, *Laughter on the Stairs,* 1953.

Other significant early arrivals were *P. zonale*, first cultivated by the Duchess of Beaufort in the early 1700s; the scented-leaved *P. capitatum* (a straggly species brought from Holland to England by Willem Bentinck in 1690 and later to become commercially important in the production of 'geranium oil'); and *P. inquinans* which Henry Compton, Bishop of London, grew in Fulham in 1714. The latter species' first appearance in print in England was in the pages of Johann Dillenius's *Hortus Elthamensis* of 1732, being a catalogue of the plants, including a sprinkling of other pelargoniums, growing in the garden of James Sherard, apothecary, at Eltham in Kent.

A softly woody plant, *P. inquinans* grows into a small shrub of one to two metres. If touched or squeezed its large, velvety leaves turn a light, rusty brown (*inquinans* means 'staining'). Found in the coastal scrub and valley bushveld of the Eastern Cape, it has striking scarlet flowers, which no doubt caught the attention of early visitors.

Having made its way to Britain, *P. inquinans* was later requisitioned by breeders to provide the brilliant red of a host of hybrids. The contribution of *P. zonale* to these takes the form of the horseshoe-shaped zonation of the leaves. These ancestral traits have been tweaked and cosseted to persist conspicuously in today's cosmopolitan and multitudinous scarlet zonals.

Another notable early cultivar-to-be was *P. peltatum*, the original ivy-leaved pelargonium. First grown in Leiden in 1700 from seed sent from the Cape by Governor Willem van der Stel, this plant has since scrambled relentlessly into and out of the windowboxes and hanging baskets of the world, many of its cultivated descendants being barely distinguishable from their Cape ancestor. The consummate flaunters of these plants are, of course, the chocolate-box chalets of Austria and Switzerland, from whose balconies and balustrades myriad ivy-leaved incarnations copiously cascade.

By the mid- to late eighteenth century, a respectable number of pelargonium species had been brought to Europe and featured in the botanical publications of the time. Nicolaas Burman described some in *Specimen botanicum de Geraniis* of 1759, calling them all geraniums. The Spanish botanist Antonio José Cavanilles detailed 71 species (a reflection of just how many had been imported to, or cultivated in, Europe by this time)

'Geranium African noctu olens', the 'African night-scented Geranium', from Jan Commelijn's *Horti Medici Amstelamodensis* of 1697. This is one of the finest of the early illustrations of *Pelargonium triste*, a species that had found its way from the Cape into French and English gardens by the mid-seventeenth century.

The word pelargonium comes from pelargos which literally means a stork, in allusion to the beak-like form of the seed pod. Similarity in the derivation of the names geranium and pelargonium – Cranesbill and Storksbill – is sufficient to indicate the reason for the confusion and ambiguity which often occur in the use of these names.

H. G. Witham Fogg, *Geraniums and Pelargoniums*, 1964.

In making choice of a name for any of our friends that we have been accustomed to speak of as geraniums, take care to look about you. If there are any botanists within hearing, say 'pelargonium,' and take all the consequences. But if none of those exacting and fastidious gentry are in the field, speak of the plants as 'geraniums,' and you will have the good fortune to be understood by the entire audience without exception.

Shirley Hibberd, *The Amateur's Greenhouse and Conservatory*, 1873.

in his *Monadelphiae classis dissertationes decem* of 1787, again including them in the genus *Geranium*.

In 1789 the botanical name *Pelargonium*, as distinct from *Geranium*, was introduced by L'Héritier who described 89 species in his *Compendium Generalogium*. An unfinished and unpublished manuscript, this work nevertheless affords him recognition as 'father' of the genus. That the name *Pelargonium* was immediately adopted by William Aiton and included in his *Hortus Kewensis*, published in the same year, is vindication of L'Héritier's authorship and chosen nomenclature, and as *Pelargonium* the plants have remained. Incidentally, 102 species of *Pelargonium* are listed in *Hortus Kewensis*, 47 of them introduced by Masson.

Many modern nurserymen and gardeners, having presumably not updated their catalogues since the late eighteenth century, still insist on the use of 'geranium' when describing pelargoniums. The former name should be strictly confined, in popular and scientific usage, to the plants of that particular genus, the cranesbills. *Geranium* itself includes many hybrids and cultivars that are hardy and attractive garden occupants, but it is *not* the same as the pelargoniums of the Cape and their multitudinous descendants!

By the beginning of the nineteenth century, pelargoniums had become well established and were proving to be of great interest to gardeners and nurserymen who had been crossing the various species, mainly out of curiosity, for half a century and more. But now the production of commercially valuable cultivars began to develop into a fully-fledged industry, with the emphasis on smaller plants with larger, more abundant blooms. Soon, many new cultivars became available in a few London nurseries, prominent among them being Colvill's which, by the 1820s, could boast as many as 500 varieties under almost 4,000 square metres of glass.

It was while employed at Colvill's that Robert Sweet gathered much of the material for his monograph *Geraniaceae*, published in five parts between 1822 and 1830. This work describes the other members of the family Geraniaceae, including *Monsonia*, *Erodium* and *Geranium*, as well as *Pelargonium*, but it is the latter that make the greatest impact and about 500 are illustrated with hand-coloured engravings, the majority taken from original paintings by Edwin Smith.

Now every garret and cottage-window is filled with numerous species of that beautiful tribe [Pelargonium], and every greenhouse glows with innumerable bulbous plants and heaths of the Cape. For all these we are principally indebted to Mr Masson.

Sir James Smith, 1812.

Sweet was a busy man. In 1826, he published his *Hortus Britannicus* in which 402 pelargoniums were listed. Just over 90 of these were species from the Cape, with a scattering from other parts of Africa and its islands, and Australia. The remaining 300 or so were hybrids.

Fiery and Scintillating

Although the professional nurseryman James Colvill was the leading hybridiser in the early days, a dedicated amateur, Sir Richard Colt Hoare, made very important contributions to the progress of the pelargonium.

Colt Hoare lived at a fine country seat at Stourhead, Wiltshire. Following the death of his young wife after only two years of marriage, he sought solace in gardening and antiquities. As far as pelargoniums were concerned, he is known to have obtained stock material from Colvill and from Lee and set about investigating the results of crossing the various species and early hybrids that he had at his disposal. Methodical and systematic, he assigned his plants to five classes: *inquinans* (scarlet), *purpureum* (purple), *maculatum* (spotted), *striatum* (streaked), and *grandiflora* (large-flowered).

As far as the scarlet group was concerned, Colt Hoare set about trying to obtain novel red-flowered varieties based on *P. fulgidum*. Four years of pollinating and planting followed before a clutch of plants of the desired colour appeared amongst his seedlings. This achievement he considered to be the 'grand epoch in the history of this plant from

Sir Richard Colt Hoare (1758–1838) immersed himself in archaeology and gardening following the death of his young wife during pregnancy. He conducted the first excavation of Stonehenge in 1798 and investigated numerous other ancient sites on Salisbury Plain. When he was not digging things up, he was planting them, and his pioneering and intensive work on pelargoniums resulted in myriad hybrids. Colt Hoare was meticulous in recording the results of his pelargonium experiments, keeping pressed specimens of the hybrids he obtained.

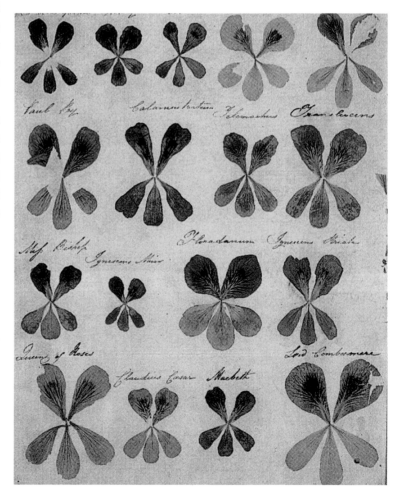

which the numerous varieties of scarlet, with its different shades have originated.' He was not being over dramatic, as one of the products that he christened 'ignescens' ('fiery'; the others were 'scintillans' and two unnamed dwarf varieties) became the basis of many red-and-black-flowered hybrids. Its genes are likely still to be coursing through the veins of many modern specimens, whilst 'ignescens' itself may persist under the guise of the cultivar 'Unique Aurore'.

Colt Hoare's *purpureum* hybrids derive most probably from the handsome species *P. cucullatum,* but the other parent, be it species or hybrid, and the hybrid parents of subsequent generations, are lost in the mists of time, as is the case with so many of this and later vintages. The fact remains, however, that Colt Hoare produced a great array of hybrids (600 by 1821) that were quickly utilised by other nurserymen. Many were also featured by Sweet in his *Geraniaceae*.

By the 1820s and '30s, pelargoniums were prominent among the plants marketed by nurserymen and florists. Each business seemed to have its own eponymous hybrids, and some species, including *P. triste,* were still offered for sale. Many amateur gardeners also found that by applying a bit of basic biology and simple techniques they could create their own varieties and easily propagate them from cuttings. The popularity of the plants increased and scores of new hybrids were reported in the numerous horticultural publications of the day.

So far, the emphasis had remained on a relatively small number of varieties, the majority not so far removed from a few Cape species and their immediate descendants. Some of the earliest arrivals out of Africa now, however, began to receive renewed attention, notably the scarlet zonals.

Tom Thumb and the Origin of Species

Although something of a late developer, given that it had been in Europe for well over 100 years, the wild *P. zonale*, with its variably horseshoe-patterned leaves, quickly made up for lost time. Once its potential and adaptability were recognised when combined with *P. inquinans*, it went on to make a prodigious contribution to horticulture. It has, however, seen as much, or more, exposure out of doors than in, and its hybrid descendants continue to crowd into urban beds and borders in summer in the way that tourists throng the streets in the same season.

Pelargoniums with coloured (i.e. not exclusively green) leaves are mentioned in Miller's *Dictionary* and in Sweet's *Hortus Britannicus*. The latter refers to it as 'marginatum'. A stripe-leaved pelargonium described by Thomas More in Robert Furber's *The Flower Garden Displayed* of 1734 is probably the same plant. Noting how he acquired it in Paris, More observes that 'The leaves … are edged with cream colour, and it makes one of the most beautiful shrubs among the greenhouse plants.'

Despite the attraction of these quirky-leaved cultivars, they received little more than passing interest for many years subsequently. Almost a century went by before efforts were made to combine the various traits to produce a plant that was fancy-leaved, robust and floriferous and, thereby, perfect for the new fashion of 'bedding out', a style of gardening that attracted as much vitriol as it did ardour.

'General Tom Thumb', which was to become one of the most widely used bedding cultivars, was first offered for sale in 1844. Apparently, the original plant was found amongst seedlings arising from the cross-pollination of the zonal 'Old Frogmore Scarlet' and an ivy-leaved pelargonium that edged the basket in which they both grew on the estate of R. Piggott of Dullington House, Newmarket. The new seedling matured to display the flower colour of the zonal but had inherited some of the low growing habit of the ivy-leaved, thereby making it ideal for bedding out.

The novelty plant was given to the young son of the house and, after a period of neglect, was consigned to the compost heap. Fortunately, it was rescued and resuscitated by the gardener, Mr Wilson, who recognised its qualities and sent a specimen to the nurseryman and pelargonium specialist, William Ayres, who introduced it to the horticultural

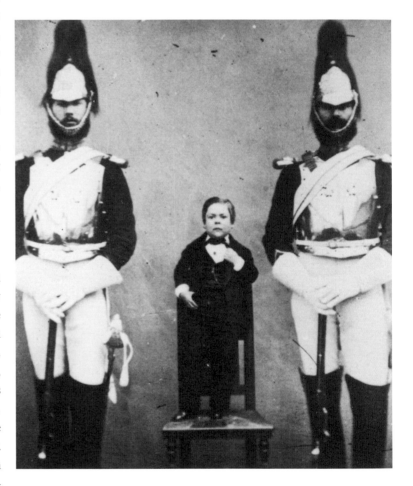

An unlikely name for a pelargonium was 'General Tom Thumb' after Charles Stratton (1838–1883), the famous dwarf who toured with Phineas T. Barnum's circus. The cultivar, which became one of the most popular of its day, was named after the 'General' because the nurseryman responsible for developing it had a promotional card in his pocket at the time. *Picture Collection, State Library of Victoria, Australia*

What are commonly termed 'Scarlet geraniums' are now so greatly improved in quantity and quality as to have become one of the most popular flowering plants for adorning the conservatory, greenhouse, dwelling-house, and flower garden. So very generally are they cultivated, and so strikingly ornamental for a greater part of the year, that we may exclaim, 'what should we do if bereft of them?' The vacuum occasioned could not be equally well filled up.

The Floricultural Cabinet, 1851.

Grieve went on to experiment with pelargoniums that had golden- and silver-edged leaves and, by all accounts, he spent long hours crossing, cutting, planting and nurturing his charges, all the while being on the lookout for the 'sports' that appeared by chance in the seed trays. When bronze and green zonals appeared on the scene they were added to the mix, allowing Grieve to breed a huge variety of cultivars over 40 years.

Grieve kept meticulous records and describes the plants that were obtained from particular crosses; not all of them had the aesthetics or vigour to be successful, but noteworthy among his achievements were the 'nosegay' varieties characterised by their large trusses of narrow-petalled blooms.

At the end of his book, Grieve details those cultivars that he considers the best of their class. Under 'Golden Variegated Zonals or Golden Tricolors' he lists just over 50 names, including a number with royal and aristocratic appellations, and one called 'Peter Grieve' which:

> gained first prize at the Special Pelargonium Show, held at South Kensington on 22nd May 1869, as the best golden tricolor introduced, thirty-four varieties competing for this much coveted prize.

In the context of variegated pelargoniums, Grieve pays particular credit to 'Mr. Beaton ... for the results of his persevering exertions in improving this now splendid class of bedding and conservatory flowers.' This was Donald Beaton, originally from Ross-shire in the north of Scotland. He later moved south to work as a gardener and experimented widely with plant hybridisation, especially pelargoniums. He was described by the Bishop of Rochester as 'Dear, quaint, old Donald Beaton' and by Charles Darwin as 'A clever fellow and a damn cocksure man'!

Amongst Darwin's almost innumerable investigations seeking evidence for his theory of evolution was an interest in the heritability of various features of hybrid pelargoniums, particularly the occasional appearance of symmetry in flowers that were naturally asymmetrical from top to bottom. This became a subject of great debate between Darwin and various of his correspondents, including Beaton. Pelargoniums may not feature prominently in *The Origin of Species*, but they certainly made a contribution to the great man's thinking.

world. A promotional card that Ayres had in the pocket at the time, advertising the appearance of 'General Tom Thumb', the celebrated dwarf in a show promoted by Phineas Barnum in London, gave the plant its name. Over its early years, it is likely that 'General Tom Thumb' was sold in millions as the Victorians stuffed their flowerbeds, private and public, with brilliant blooms.

The pre-eminent breeder of zonals in the nineteenth century was Peter Grieve, and his little book entitled *A History of Ornamental-foliaged Pelargoniums with practical hints for their production, propagation, and cultivation*, published in 1868, provides a quaint insight into his special-ised world and interest.

Grieve was born at Allanton in Berwickshire in 1812, and became gardener to the Earl of Lanesborough before moving into the employ of the Rev. Benyon at Culford Hall, Bury St Edmunds, in 1847. Here he became interested in crossing zonal-leaved pelargoniums, recalling that 'in 1853 or 1854 ... my attention was first directed to this subject.' He began by cross-pollinating 'Flower of the Day' with 'General Tom Thumb', the result being christened 'Culford Beauty'.

Pelargonium 'Dolly Varden'. This cheery, fancy-leaved zonal is named after the coquettish character in Charles Dickens' *Barnaby Rudge*. Dickens loved flowers, but 'scarlet geraniums were his favourite of all', according to his daughter. His buttonhole and two beds in his garden at Gad's Hill, Kent, were occupied by scarlet pelargoniums whenever possible. When he died in 1870 his coffin was adorned with a wreath of them.

below
Dolly Varden as imagined by the celebrated Victorian artist William Frith (1819–1909). A friend of Dickens, Frith managed to balance a busy professional life with a wife and 12 children and a mistress with seven more.
©*Victoria and Albert Museum, London*

Wilde malva *Pelargonium cucullatum* in bloom on the Cape Peninsula. One of the tallest shrub pelargoniums, this species was planted in Cape Town gardens in the nineteenth century, a rare ornamental use of a native plant when English garden flowers were de rigueur.

In making a short cut over a stony hill, covered with low bushes, I noticed, in blossom, in the fissure of a rock, the elegant Pelargonium tricolor. This was like recognising an old forgotten acquaintance, of a pleasant character; for the existence of this old, but elegant and delicate inhabitant of English green-houses had quite passed from my mind, till scarcely raised above the stone on which it grew, a large cluster of its pure-white blossoms, shaded in to blackish crimson, met my eye, in this inhospitable region, and revived many associations in connexion with the persons under whose care I had seen it cultivated.

James Backhouse, *A Narrative of a Visit to the Mauritius and South Africa*, 1844.

Wilde Malva

Although Darwin later displayed some interest in the Cape flora, he didn't have a great deal to say about it at the time of his brief visit there in the *Beagle* in 1836. He certainly made no mention of pelargoniums, although he must have encountered one or two.

The profusion of pelargoniums that smothered the low, seaward slopes of the Cape of Good Hope Nature Reserve one particular spring is one of the most memorable sights that we enjoyed at the Cape. There were great masses of them, pink cloudy banks of thousands of plants. This great botanical blush was made up of *P. cucullatum*, known locally as wilde malva, their seed having germinated *en masse* following a fire that swept through the fynbos the previous autumn, leaving a perfect seedbed. Slight variations in the strength of pink in the petals, the occasional white sport, and the odd natural hybrid with *P. betulinum* (birch-leaved pelargonium; my favourite), gave little indication of the super-abundance of cultivars that were obtained from this great-grandfather of pelargoniums.

P. cucullatum was an early arrival in Holland from the Cape and was probably another of the species in Willem Bentinck's luggage when he came to England in 1690; it could certainly be found growing at the Chelsea Physic Garden by 1724.

The importance of *P. cucullatum* from the horticultural point of view lies in its standing as the principal ancestor of the regal pelargoniums, also known as French, Lady or Martha Washington, fancy, show or, in semi-official horticultural parlance, '*P. x domesticum*'.

It is likely that the first cross to produce what we now know as regal pelargoniums was between *P. cucullatum* and *P. grandiflorum*, a species

from the high western Cape mountains sent to Kew by Masson in 1794. A subsequent cross with *P. fulgidum* is thought to have introduced bright red into the palette, while *P. betulinum* chipped in a deeper pink flower and a smaller leaf. Thereafter, the original species made little or no further input while the various cultivars were back-crossed with each other to produce new varieties, and a keen eye was kept on the seedlings for any unusual sports.

In their early days, these crosses were known as 'show', 'fancy' or 'large-flowered' pelargoniums. Their sizeable, cheery blooms made them very popular and they were widely exhibited at the horticultural shows that were all the rage in Victorian days.

It was not until late into the nineteenth century that the word 'regal' was introduced to describe these showy plants. One source claims that they were so-called because they had their origins on the royal estate at Sandringham. Another suggests that it was coined by William Bull, nurseryman on the King's Road, Chelsea, as a marketing ploy to appeal to the royalist devotions and affections of his customers and the population at large. Some of the named varieties also reflected this prevailing sentiment – 'Queen Victoria' was one, 'Princess of Wales' another.

A multitude of highly decorative regals was now grown on an industrial scale by market florists. Being longer-lasting than a bunch of cut-flowers and not much more difficult to maintain, they helped satisfy the demand for colourful, convenient and practical blooms for Victorian drawing rooms, windowsills and doorsteps.

Over the years, regals have been bred to produce showy flowers of every shade of pink, and from a deep mahogany- or maroon-purple

Rooimalva ('red malva') *Pelargonium fulgidum* is likely to have introduced a splash of scarlet into the early regal cultivars. Found on the Cape west coast, this species flowers from mid-winter to spring and can reach a height of 2 m as it clambers among convenient shrubs.

that could just about pass for black, through to pure white. The petals have become increasingly veined, striped, spotted or blotched with other shades, and many have incised or fringed edges. The majority remain, to me at least, plush but dignified; others, through a combination of fluffy pink and frills, have advanced (I use the word advisedly) to that extreme of kitsch traditionally devoted to hand-knitted toilet-roll covers masquerading as fictional princesses or poodles in waistcoats. For all that, these remain strangely appealing and, in a demonstration of gross want of refinement, I have been known to sit 'Mendip Barbie' between 'Lord Bute' and 'Montague Garibaldi Smith'. The bottom line, however, is that regals provide wonderful and long-lasting colour and are a joy to grow.

Scents and Sensibility

Although regals are crossed mainly within their group to produce yet more variety, they have occasionally broken away to make important contributions in other departments. One such cross was with *P. graveolens* to give a scented-leaved plant with showy flowers.

The leaves of most pelargoniums have at least some scent. The mere touch of a leaf can be redolent of a greenhouse on a hot summer's day, evoking happy hours ensconced therein in the company of aphids. Some species, however, have particularly distinct aromas, containing volatile oils that give them a characteristic fragrance. Over 70 different

The angel pelargonium 'Cottenham Surprise' is one of a series of beautiful and floriferous pelargonium cultivars bred by Mervyn Haird of Cottenham, Cambridge. The result of crosses between *Pelargonium crispum* and regals made originally in the 1930s by a London schoolmaster, Arthur Langley-Smith (?–1953), angels have made a relatively recent but impressive entrance onto the pelargonium stage.

If the majority of Pelargoniums are deficient in fragrance, nature has made up for that apparent deficiency, by the splendour of the blossoms; and, as it were, to equalise her gifts, certain kinds whose flowers are less showy, nay, even a dingy hue, have a delightful perfume; some during the evening and night, and others when rubbed against, or when the wind lashes the leaves and branches against each other.

Charles McIntosh, *The Greenhouse, Hot House and Stove*, 1838.

At this point of the story the subject becomes too large to be handled on the present occasion, and I shall say nothing of the Tricolours and Bicolours and Ivy-leaves and the Uniques, that during a period of about ten years overran all the gardens, furnished a common theme for conversation at every table, supplied all the business men and a few peers of the realm with buttonhole flowers, persuaded half the human race that providence had designed gardens for one tribe of plants alone, and that possibly the pelargonium itself was the tree of knowledge of good and evil that stood in the midst of the Garden of Eden for the delight and perplexity of the very first gardeners.

The Gardeners' Chronicle, July 1880.

oils have been identified and, in the wild, these serve to cut down water loss and to deter browsers such as insects and antelopes.

The blooms of many scented-leaved species are relatively demure. Their leaves, however, make up for this deficiency, if such it is, by being powerfully scented, especially when touched or bruised. The scent varies between species and, while the experts may not have the imaginative vocabulary of wine tasters, they have assigned the various species and cultivars to variations of, amongst others, camphor, pine, cedar, balsam, verbena, lavender, musk, cinnamon, ginger, blackcurrant, strawberry, turpentine, and any citrus fruit or nut you care to think of. The most familiar are rose in *P. capitatum* and *P. graveolens*, apple in *P. odoratissimum*, balm in *P. quercifolium*, mint in *P. tomentosum*, and billygoat in *P. papillionaceum*. With the possible exception of the last-named, these species or their first-generation hybrids were widely grown in the home in the eighteenth and nineteenth centuries purely for their pleasant scents, and they remained largely untampered with by the enthusiasts who were obsessively squeezing every last drop of variety out of the other groups. A few early generation scented-leaved cultivars did become popular, however, not least the variegated-leaved, delicately-pink-flowered 'Lady Plymouth' (a hybrid of *P. graveolens*), and the ubiquitous small, oval-leaved '*P. Fragrans*' with its profusion of small, white flowers.

More scented hybrids are coming onto the market nowadays, some of them fairly impressive in both the floral and aromatic departments. The peppermint ones (there is now even a chocolate-mint variety) are perhaps the most unexpected, associating mint as we do with that garden herb, but these pelargoniums, with their soft, hairy leaves, are quite remarkably, and naturally, minty. They are vigorous growers and require plenty of space to clamber around as well as occasional disciplining with the secateurs.

The scented pelargoniums are valuable not just as ornamentals but also as the source of 'rose-geranium', an essential oil used on its own in cosmetics, pot-pourri and aromatherapy or, more often, to supplement more expensive rose oils in these products. A number of species have been crossed to deliver these oils, the major contributors being *P. graveolens* and *P. capitatum*. These and their hybrids are grown on a large scale in places such as Réunion Island in the Indian Ocean, in countries round the Mediterranean and other warm-temperate areas.

Angel Delight

The scented-leaved pelargoniums have contributed to one of the relatively recent additions to the major groups of cultivars. When *P. crispum*, a very pretty, lemon-tangy species from the south-western Cape, was crossed with a number of regals, initially with one called 'The Shah', the product resembled a miniature regal. The first of the subsequent line to be introduced to the wider world, in 1935, was 'Catford Belle'. It and its contemporaries were known as Langley-Smith hybrids, after their creator, the eponymous Arthur, a schoolmaster from Catford, south London.

In 1958, Derek Clifford, a pelargonium expert, felt that these cultivars, some retaining the crinkly leaves and citrus scent of *P. crispum*, were sufficiently distinct to qualify as in independent group, which he christened 'angel', because they resembled a pelargonium called 'Angelina' described by Henry Andrews in 1792. This may have been a cross between *P. crispum* and *P. betulinum*.

The angel group of pelargoniums has now become firmly established and its members are certainly extremely attractive. The plants are typically, but not exclusively, quite small (or can be kept so) and compact, and flower as if their lives depend on it (which they do, actually). Specialist nurserymen have produced lines such as 'Cottenham' and 'Quantock', which successfully retain the combination of charm, resilience and the evocative scent that characterises many of their wild ancestors.

This scented-leaved cultivar is categorised by pelargonium expert John Cross as 'P Crispum Variegatum (Variegated Prince Rupert), single, lavender, carmine veined. Leaf: Small crenate, lemon-scented, Lavender green 000761 – border, Primrose Yellow 601/03.' The numbers refer to the colours on the Horticultural Colour charts issued by the British Colour Council and the Royal Horticultural Society.

This Good-tempered Flower

The Victorian age, as we have seen, was a wonderful one for pelargoniums in Britain, and even at the beginning of the twentieth century 'no garden or greenhouse was complete without a collection of this good-tempered flower', as pelargonium enthusiast and writer John Cross (founder, in 1951, of the British Geranium Society, later to become the British Pelargonium and Geranium Society) put it. Many hundreds, nay thousands, of cultivars had been produced and, although often short-lived thanks to the vagaries of fashion, they were replaced with a constant stream of new ones.

Sadly, the golden age of the pelargonium finally crumbled under the effects of two world wars, when cultivars were lost as greenhouses became neglected and too expensive to heat (and, anyway, it was forbidden to grow ornamentals under glass in the First World War) and gardens were turned over to food production. Gardeners and nurserymen enlisted in the army and were often not replaced, and the fashion for 'natural' planting, as espoused by gardeners such as William Robinson and Gertrude Jekyll, bred antipathy to bedding plants.

After the Second World War and years of drab living and rationing, gardening enjoyed a revival in Britain. Pelargoniums also enjoyed a surge of popularity, helped along by expert horticulturists and communicators such as W. A. R. Clifton, Derek Clifford, and John Cross. At the Chelsea Show of 1949, garden writer Miss Eleanour Sinclair Rohde is quoted (by Cross) as declaring at Clifton's display:

> I was delighted to find those Victorian favourites, geraniums, in a remarkable range of colours. Geraniums will, I think, once more become a cult. Big bowls of them in mixed colours are so old-fashioned that they appear ultra-modern, and no other flowers light up with such remarkable brilliance in artificial light.

left

The ivy-leaved or trailing pelargonium is descended from *Pelargonium peltatum*, a species with flowers varying from white to deep pink. The wild plant is a bit straggly, and it remained largely unchanged for many years after its introduction in the early eighteenth century. From about the 1870s the breeders, prominent among them Victor Lemoine, succeeded in making it more robust and souped-up its flowers. Today's varieties have leaves that can be variegated, meshed or otherwise patterned, and double blooms with scarlet candy stripes. Dark margins around pale-centred petals are, as in some tulips, the consequence of otherwise benign viral infection. *Anthony Hitchcock/South African National Biodiversity Institute*

above

A selection of pelargonium cultivars demonstrating the variety and range of flowers and foliage that have been developed from just a handful of wild species.

1 'Fir Trees Mark' (Regal)
2 'Vancouver Centennial' (Stellar)
3 'Sarah Don' (Angel)
4 'Berkswell Pixie' (Angel)
5 'Ursula's Choice' (Angel)
6 'St Elmo's Fire' (Stellar)
7 'Apple Blossom Rosebud' (Rosebud)

8 'Quantock Cobwebs' (Angel)
9 'Lord Bute' (Regal)
10 P. quercifolium cultivar (Oakleaf)
11 'Fringed Aztec' (Regal)
12 'Concolor Lace' (Scented-leaved)

left
The pelargonium cultivar 'Renate Parsley' commemorates a well-known South African horticulturist. Its delightful and prolific flowers more than make up for its rather short stature and slow growth. It retains all of the charm of its parents – *P. tricolor*, first collected by Masson in 1791 and which contributes the bicolored flowers, and *P. ovale* which has oval, grey-green leaves.

above
A colourful mix of zonal, regal and stellar pelargoniums. 'Bird Dancer' hovers above an anonymous regal, flanked by 'Red Witch' on the right.

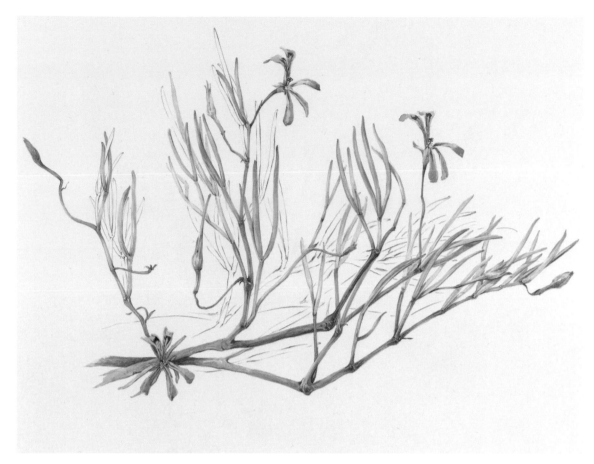

Interest in the plants has continued to burgeon following their restoration to the windowsill in the 1950s. Developments overseas introduced a number of new lines including Formosa from North America in the 1960s and stellars from Australia in the 1970s.

The last few decades have also seen the appearance and, in some cases, reappearance, of tulip-flowered, harlequin- and paintbox-flowered, pansy-faced, and lace-leaved hybrids. A yellow pelargonium, seemingly the 'Holy Grail' of the breeders, was produced in 2006. Blue is also much sought after by the breeders, and a few varieties boasting an epithet of this colour are now available. Calling them blue doesn't make them blue, of course, and it's often as much a question of semantics as chromatics, as one man's 'blue' is another's 'purple'.

In recent years specialist growers such as Helen and Mark Bainbridge (Fir Trees Pelargoniums), Hazel and Dick Key (Fibrex Nursery, and holders of the National Collection of Pelargoniums), Ken Dymond, Tony Burgess, and Brian Dixon, have produced a wealth of interesting and attractive new cultivars, which go some way towards counterbalancing the industrial-scale production of cheery but commonplace varieties by the big commercial concerns. But these, too, have their place in pots, tubs, hanging baskets, and window-boxes, and in the regimented plantings in municipal parks and gardens that so gladdened the hearts of the Victorians.

Nemesias from the Cape have been in cultivation for over 300 years. While the original species are very attractive, the urge to improve on these, or at least vary their flower colour or growth form, has resulted in an abundance of hybrids for every corner of the garden. The orange one is 'Prince of Orange'. No surprises there, then.

CHAPTER TWELVE

Bear's Ears and Garden Diamonds

The 'Golden Age' of plant collecting at the Cape was, when all is said and done, less of an age and more of a moment. Despite such exploratory transience, however, many of the plants that were collected during this frenetic period had enough staying-power to find and retain, in one form or another, a place in our gardens and greenhouses.

Although the bulbs and pelargoniums were most prominent, a diverse bunch of other Cape plants has also succeeded in carving a solid horticultural niche over the years and, unlike the ericas and proteas, has withstood the vagaries of fashion and technology. Indeed, those plants described in the preceding chapters account for only a fraction of the almost 120 genera of Cape plants in cultivation that are listed in the various gardening reference books. Some species and their cultivars have even attained levels of popularity to rival the pelargoniums. Prominent among these is *Streptocarpus*.

Bowie's Bequest

Sometimes unnecessarily referred to in the horticultural trade as the Cape primrose or wild gloxinia, members of the genus *Streptocarpus* are found in eastern Africa from Ethiopia in the north down to the Cape, with a few species in Madagascar and Angola. Two of the 40 or so southern African species occur in the Cape Floral Kingdom. These are *S. rexii* and *S. meyeri*.

S. rexii grows on the shady, wooded slopes of the southern and eastern Cape. Seed collected near Knysna by James Bowie was sent to Kew in 1826 and the first plants bloomed there the following year. For almost three decades *S. rexii* flowered in glorious isolation at Kew and only a two-flowered variety derived from the species and named, sensibly if unimaginatively, '*S. biflorus*' made its way via the nurserymen into limited commercial cultivation.

In 1855, two more species – *S. gardenii* and *S. polyanthus*, were introduced to Europe. The latter, confusingly, was sometimes given the name of '*S. rexii* var. *S. biflorus*'. The first recorded hybrid *Streptocarpus*, achieved by the French horticulturist Victor Lemoine, was supposedly between '*S. biflorus*' and *S. polyanthus*, and in 1859 the offspring of this cross were labelled '*S. bifloro-polyanthus*', even though contemporary illustrations suggest that it was *S. gardenii* and not '*S. biflorus*' that was one of the parents. In 1862 Lemoine included this hybrid and five selected strains in his catalogue, but the original plants were lost.

Unable to recreate his original hybrids, Lemoine created a new cross between *S. rexii* and *S. polyanthus*. He offered it for sale in 1882, but in 1889 a nurseryman in Poitiers by the name of Bruant achieved the same cross and called his plants '*S. x bruantii*'. This did not please Lemoine, and he requested that the name '*S rexii-polyanthus*', which he had bestowed on his hybrid, should be retained. '*S. x bruantii*' had, however, been published in the *Revue Horticole*, so its editors tried to pacify all parties by renaming it '*S. x controversus*'! The laws of horticultural precedence dictate that the first officially published name is the one that stands, and '*S. x bruantii*' remains valid. Be this as it may, the French hybrids languished and Lemoine, who died in 1911, remains less well known for his streptocarpus than for his numerous lilac cultivars. He was awarded the RHS Victoria Medal of Horticulture, the first foreigner to be so honoured.

New streptocarpus species, meantime, had been slow to appear but one, *S. saundersii* from KwaZulu Natal, was introduced into cultivation in 1861, and in 1876 a cross between it and *S. rexii* was christened '*S. x greenii*'. Things gathered pace after 1882 when the nurseryman J. H. Veitch crossed newcomer *S. parviflorus* from the Transvaal with *S. rexii* to produce plants subsequently known as Veitch's Original Hybrids.

Further momentum built up with another arrival, *S. dunnii*, also from the Transvaal. This species first flowered at Kew in 1886 and is unusual in that its blooms range in colour from 'old rose' to 'deep crimson', a striking contrast to the traditional white and shades of blue and mauve that characterised those species known so far. Joseph Hooker was greatly impressed,

below

Streptocarpus rexii was the first strepto-carpus to be discovered (by James Bowie in 1818). The species is named after George Rex (1765–1839), once rumoured to be the love child of Prince George (later George III) and his 'wife' Hannah Lightfoot. Sadly for historical romantics, this has been proven to be untrue. A prominent figure in Cape life, Rex was a welcoming and sociable host to travellers, including plant collectors, at his property Melkhoutkraal at Knysna, the settlement he founded and whose growth was boosted with his own contribution of 13 children. His woodland plant occurs east along the Cape coast from George and north to KwaZulu Natal.

[Streptocarpus] colours are most diverse, ranging from pure white, bluish delicate rose-pink, to magenta and maroon; also from the palest lavenders and mauves to the richest violet-purples imaginable, browns and terra-cottas in various shades being also represented. The beauty of the flowers is further enhanced by the chaste clear pencillings and bold contrasting markings in the throats of the majority; so that altogether it may be said of the Streptocarpus that it has given some of the most dainty floral gems to our greenhouses during the summer.

John Wright, *The Flower Growers Guide*, 1897.

left
The great plant breeder Victor Lemoine (1823–1911) developed some of the earliest streptocarpus hybrids and a multitude of other cultivars, including gladioli and the first double-flowered zonal pelargonium, in his nursery at Nancy, France.

right
A typically dazzling display of streptocarpus cultivars by Dibley's Nursery at 'Gardening Scotland'.

calling it 'quite the monarch of its beautiful genus', and named it after Edward Dunn, the geologist who had sent him the first seeds in 1884.

Despite its unusual and attractive flowers, *S. dunnii*'s relatively simple structure and foliage (it belongs to the unifoliate group of *Streptocarpus*) didn't endear it to the florists. In 1887 or thereabouts, however, William Watson, assistant curator at Kew, crossed it with *S. rexii* and *S. parviflorus* (which have more attractive growth forms, notably a basal rosette of leaves) to produce, respectively, 'S. x *kewensis*' and 'S. x *watsonii*'. The following year *S. dunnii* was relegated to the garden bench, while work continued on the hybrids.

This represented a milestone in streptocarpus development and, as John Wright, author of the *Flower Grower's Guide*, observed in the late 1890s, 'It may be presumed that even Mr Watson could scarcely have foreseen that his two little floral children would soon bring in their train the marvellously diversified and beautiful varieties now so easily obtained from seed.'

The work was continued by John Heal of Veitch's nursery, who had seen 'S. x *kewensis*' and 'S. x *watsonii*' in the succulent house at Kew and managed to procure a few specimens on which to experiment. 'The result', according to Wright, 'exceeded the expectations of every one who takes an interest in these flowers.' He records how the display of 'Veitch's Hybrids' at the RHS show in the Temple Gardens, London, in 1897 (the year of the Diamond Jubilee) 'excited expressions of admiration and surprise' with its flowers of 'pure whites without any trace of colouring, then white with dark blotches, magentas with rich and distinct shades, violet, purple, rose, pink, and various tints practically indescribable.'

By the beginning of the twentieth century a respectable range of streptocarpus cultivars was available, and backcrossing with *S. rexii* was undertaken when necessary to retain the desirable foliage and growth form of that ancestral Cape plant.

For a few years, however, activity was relatively subdued, until in 1914–16 Richard Lynch, curator of the Cambridge Botanic Garden, experimented with crossing *S. cyaneus* (introduced in 1907 from

Swaziland) and *S. denticulatus* (1912, Eastern Transvaal) to produce more hybrids. Efforts over the next 20 years then reverted to improving the quality of the so-called '*S. x hybridus*' plants, being strains resulting from the crossing of a number of different species with *S. rexii*.

In the 1930s, work by W. J. C. Lawrence at the John Innes Institute at Merton, London, on the inheritance of flower colour in streptocarpus resulted in the production of two significant new hybrids: 'Merton Giant' (from *S. grandis* x '*S. x hybridus*') and 'Constant Nymph' ('*S. x hybridus* Merton Blue' x *S. johannis*).

'Constant Nymph', multiflowered and 'flat-faced', has become the basis of many of the modern cultivars. A white sport that arose from it in the Netherlands was named 'Maassen's White' and zapping it and its progenitor with X-rays gave rise to further mutations. Additional crossing of 'Constant Nymph' with *S. johannis* and various colour varieties of '*S. x hybridus*' by Gavin Brown of the John Innes Institute saw the shape and form of 'Constant Nymph' retained but the flowers taking on a variety of startling colours, enhanced by variable, and often striking, markings.

With this rich resource to draw upon, a multitude of new strains of streptocarpus has been successfully developed by many breeders and nurseries, pre-eminent among today's being Dibley's of North Wales.

Streptocarpus have become a universal favourite since their appearance on the horticultural scene, and retain a charm that might have been lost as its cultivars became increasingly floriferous, flamboyant and colourful. Perhaps James Bowie would have been more a more cheerful chap if he had known just what he was presenting to the world when he sent the first seeds back to Kew almost 200 years ago.

The seeds of another Cape species that was to become almost as popular as Bowie's streptocarpus owe their introduction to Hildagonda Duckitt, of chincherinchee fame.

Hildagonda and the Reading Connection

One of a family of ten, Hildagonda was a characterful lady who lived for much of her life at the family farm, Groot Poste, near the west-coast village of Darling, famous for its wildflowers. At the age of 19 Hildagonda was engaged to a sea captain, William Brown, but the engagement was broken off after seven years. After this disappointment, Hildagonda devoted herself to looking after her bachelor brother, her nieces and the farm, to gardening, and to writing books of recipes and domestic wisdom (in her later years she was described as South Africa's 'Mrs Beaton'.)

From the gardener's point of view, Hildagonda should be thanked for bringing a level of colour and cheerfulness into our lives that she was sadly denied in hers. Her gift to gardens was a snapdragon, *Nemesia strumosa*, that can be found growing wild on the sandy coastal plain at Darling.

The original wild *Nemesia strumosa* from Bokbaai, near Darling, occur naturally in a variety of colours and are easy to grow, so very little needed to be done to prepare them for the flowerbeds of England.

The flowers of *N. strumosa* are naturally very variable in colour, ranging from white to yellow, orange and red. Often they are bicoloured, or marked with spots or blotches. It was no surprise that Hildagonda was attracted by these lovely spring blooms and her biographer, Mary Melk Kuttel, records that when gathering seeds for her garden she went to 'particular trouble to collect the different shades – rose-coloured from the coast ... scarlet from the banks of the Berg River ... as well as the more common yellow and cream.' She sent some of her seeds to Suttons in Reading, England, and the rest, as they say, is history.

In 1892, the flowers were exhibited at the RHS show under the name of '*Nemesia suttoni*', receiving an Award of Merit. Mrs Kuttel records:

> They created a furore: it was the first time these flowers, now such universal garden subjects, were seen in England...Hilda was justifiably annoyed that Sutton's labelled them *Nemesia suttonnii* (*sic*). It is a botanical custom to name a flower after its introducer to civilized circles, and the more correct name might have been *Nemesia duckittii*, or, perhaps, *Nemesia hildasiensis*.

By 1894 Sutton's had enough seeds of '*Nemesia suttoni*' to distribute commercially; Hildagonda's flowers were given high billing in the company's catalogues of the early twentieth century and have featured prominently in one form or another ever since, becoming one of the most popular annuals in the world. A remarkable legacy of a singular lady.

In her later years Hildagonda moved with her sister, Bess, to the Cape Town suburb of Wynberg. Here she lived in 'Nemesia Cottage' and took great pride in the colourful beds of wildflowers including, of course, nemesias, in which she found some distraction from the debilitating and ultimately fatal illness that blighted her last years. She died in 1905 aged 66.

Although Hildagonda Duckitt can take credit for introducing nemesias to the gardeners of the world, it is only fair to say that hers were not the first to leave Cape shores. Sweet lists three species of *Nemesia* in cultivation in Britain, namely '*chamaedryfolia*' (*macrocarpa*, introduced in 1787), '*bicorne*' (*bicornis*, 1774), and '*foetens*' (*fruticans*, 1798).

Although often labelled as such, many of the new cultivars are not variations on the theme of *N. strumosa*, despite its natural variability, but are crosses with other species, including blue- and/or white-flowered ones. Given that there are almost 30 species in the Cape Floral Kingdom, and double that number in southern Africa as a whole, it is surprising that so few others have yet made their way into our gardens. A relatively recent arrival is the coconut-scented and fried-egg-lookalike *N. cheiranthus*. It will doubtless not be the last of its clan to make its way from Cape to cultivation. Meanwhile, amongst the mass of hybrids and cultivars that now adorn our gardens, we might hope that one called 'Hildagonda' might someday find a position.

Great-flowered Daisies

In spring, particularly following a wet winter, the wild nemesias along Hildagonda's beloved west coast can put on a lovely show, albeit scattered thinly across the veld. If it's pure unadulterated Cape floral extravaganza that you want to emulate in your garden, however, then the daisies will do it for you. With over 1,000 species, the daisy family

In September the ground is literally carpeted with endless varieties of gazanias – local name, Gousblom. All shades of yellow and orange, and a porcelain-looking white one with outer petals tinted white blue. These commonest of flowers are nevertheless very pretty. Then we have the endless varieties of mesembryanthemums, from the tiniest little ones creeping along the ground to the gorgeous scarlet and purples ... As a friend of ours just arrived from England said, 'One cannot believe that all these flowers are wild, but think they must have strayed out of some one's conservatory.'

Hildagonda Duckitt, *Hilda's Diary of a Cape Housekeeper*, 1902.

Nemesia 'KLM'. This popular, long-flowering bedding plant was bred in Holland and takes its name from that country's national airline, whose corporate livery is blue and white. The flower has so far managed to retain its original colours following KLM's merger with Air France in 2004.

left

With 30 other *Nemesia* species in the Cape and 30 more elsewhere in southern Africa, more species and cultivars will inevitably soon join the ranks of *N. strumosa* in our gardens. Among the most attractive wild species are *N. versicolor* (highly variable, but sky blue in this illustration), *N. cheiranthus* (long-horned yellow and white), *N. azurea* (purple), and *N. ligulata* (yellow and white). The quirky, coconut-scented *N. cheiranthus* has already arrived in the guise of cultivars very similar to the species, such as 'Shooting Stars' and 'Masquerade'.

opposite

A group of Cape plants painted *c.*1705. The work is attributed to Henrietta London whose father, George, was gardener to Bishop Compton, and William and Mary. An inscription on the reverse, however, states 'Drawn by Mrs. London', in which case George's wife would have been the artist. The subjects are identical to those in a Commelijn manuscript of *c.*1690, based on the *Codex Witsenii*. As the latter was the source of the illustrations in the *Codex Comptoniana*, then presumably 'Mrs London' copied them from that volume when her husband was in the employ of the Bishop. The plants are, left to right, *Nemesia bicornis*, *Cotyledon orbiculata*, and *Heliophila coronopifolia*. *Badminton Archives, by kind permission of His Grace the Duke of Beaufort*

below

The spoon-shaped florets of the likes of 'Whirligig' and 'Serenity' have added an additional quirk to the conventional 'Cape daisies'. For a plant from warm Cape climes, *Osteospermum* are surprisingly hardy and can even survive a harsh Scottish winter. The numerous cultivars display a range of flower colour, from white to deepest purple and a variety of pastel shades, including the peachy 'Sunny Serena'. Also in this border is *Gazania* Christopher Lloyd (pink with turquoise centre).

(Asteraceae) is the largest in the region. While not all of its members are conspicuous and showy, many of them are spectacularly abundant and colourful and can put on a retina-searing vernal display when the sun shines. (Like the familiar Daisy *Bellis perennis* of garden lawns, the flowers have strict opening times, and remain closed when the skies are overcast).

A daisy that has made a big splash in gardens is *Gazania*, of which there are about a dozen species in the Cape Floral Kingdom. The first to be introduced to Europe was *G. rigens* in 1755. The specific name translates as 'great flowered', but this was evidently not sufficiently superlative for the nurserymen, and the first cultivars developed about 100 years later were named '*splendens*'. A parent of many hybrids is *G. krebsiana* (known originally as *G. pavonia*) with brilliant yellow to orange flowers.

Today's *Gazania* cultivars fall into a number of groups or series, including 'Chansonette', 'Daybreak', 'Mini-star' and 'Talent'. A relatively recent and rather cunning introduction is 'Sunbathers' whose semi-double flowers can't close under the cloudy conditions that would see their congeners clam up. As the horticultural equivalent of sleep-deprivation, I have to say I find this genetic infliction curiously unsettling.

Blue Streak and Nairobi Purple

A daisy enjoying a recent upsurge in garden popularity is *Osteospermum*. Formerly included in the genus *Dimorphotheca* (and sometimes still sold as such), 30 species occur in the Cape flora. Often referred to simply as 'Cape daisies', they have established themselves in today's gardens in the form of a multitude of cheerful hybrids. The hardier ones are likely to be descended from *O. jucundum*, a montane species from north-eastern South Africa, while the Cape Floral Kingdom has contributed *O. ecklonis* and *O. fruticosum* to the breeders' palette, the latter having apparently been cultivated in England since 1740.

To find an unadulterated species in cultivation today would be unusual, as the great majority now grown are the results of crossing (which they do readily). A proliferation of osteospermums over the last few decades has arisen from the original development of a couple of cultivars, one called 'Blue Streak' and the other '*prostratum*' that was most probably a variety of *O. ecklonis* (which itself is sometimes referred to as *O. caulescens*). In 1970 a variety called 'Nairobi Purple' was introduced by the RHS and also grew at Tresco Abbey in the Scilly Islands. In the 1970s, Roy Cheek planted 'Nairobi Purple' alongside '*prostratum*' at Cannington College in Somerset. These crossed quite happily and the resultant seed produced a variety that was subsequently named 'Cannington Roy'. A second suite of hybrids was produced at Cannington in 1984.

Meanwhile, a quirky cultivar called 'Tauranga' was bred in New Zealand in 1979. This has peculiar spoon-shaped ray-florets and, when brought across to Europe a couple of years later, it was renamed 'Whirligig'. The original had white, purple-centred blooms, but pink and purple ones were soon developed. A range of such 'spoon-petalled' varieties is now grown in industrial quantities for the pot-plant and bedding markets. These and other important cultivars, each representing a major component of the osteospermum market and accounting for well over 200 named varieties, have recently been developed in various parts of the world.

It should come as no surprise that osteospermums have attained such a high level of interest from breeders and popularity among gardeners in a relatively short time. They are cheery, colourful plants that produce masses of flowers over a long season in a variety of colours, from soft pastel pinks and peach to garish hues of purple. They are also surprisingly resilient, and although often described as 'semi-hardy', many will tolerate what might be considered unsuitable and unforgiving conditions. They are, however, dedicated sun worshippers, opening fully only under the brightest skies.

Garden Diamonds

The sunshine rule applies equally strictly to *Dorotheanthus bellidiformis*. This is commonly known as the Livingstone daisy (I have no idea why) or ice plant because of the sparkly papillae or bladder cells on its leaves that glisten like diminutive diamonds. The generic name translates as 'Dorothy's flower' and was conferred by German botanist Gustav Schwantes in honour of his mother.

Confusingly, the species is often referred to in nurseries by a number of archaic, obsolete or inaccurate names, most commonly *Mesembryanthemum bellidiforme* or *M. criniflorum*. It's also not a daisy in the strictest sense, being a member not of that family (Asteraceae) but of Aizoaceae. Formerly known as Mesembryanthemaceae, this has led to

Arctotis aspera from *The Botanical Register* of 1815. 'Native of the Cape of Good Hope. Cultivated here before 1710. The drawing was made at Mr. Rolls's nursery, the King's Road, Chelsea.' Although there were at least 13 species in cultivation in Vienna at the end of the eighteenth century, 16 listed in the *Hortus Kewensis* of 1813, and it was praised as a 'hardy greenhouse plant' in 1820,

Arctotis had all but vanished by the end of the nineteenth century. Only a few of the more than 30 species in the Cape Floral Kingdom are grown in gardens today. About 20 hybrids can be found if you hunt around, some under the old-fashioned label of 'x *venidio-arctotis*', although I'm not sure how much effort I'd devote to tracking down one called 'Prostrate Raspberry'.

top
Molly Ryan enjoys the glowing *Arctotis* (gousblom) in the West Coast National Park, where the displays of these daisies and other spring flowers can be breathtaking.
Peter Ryan

above
Livingstone daisies *Dorotheanthus bellidiformis* in a Scottish garden somehow manage to display the same sunny brilliance and colourful variety as their wild ancestors.

the name 'mesemb' being applied liberally to the almost 700 fleshy-leaved species that comprise the family.

A dwarf, succulent annual (a rare combination) in the Cape, the Livingstone Daisy is popularly known there as Bokbaaivygie, meaning Buck Bay mesemb, after that spot on the west coast where it grows prolifically. Its variety of different shades of white, orange, pink or red blooms are every bit as colourful as the cultivars descended from the species.

De l'Obel's Border Patrol

Fortunately, there are plenty of Cape plants that are less dependant on the vagaries of the northern sunshine than the daisies and mesembs to perform at their best. One such resilient flower takes us back to the early days of Cape exploration and commemorates the name Mathais de l'Obel, the French botanist who, in 1605, was the first to mention a Cape plant collector (Gouarus de Keyser) and featured two of his plants in print.

Part of the bluebell family, *Lobelia* is a cosmopolitan genus of about 300 species ranging from tiny annuals to shrubs. Those most commonly cultivated today are the statuesque North American species with bright red flowers, of which the Cardinal Flower *L. cardinalis,* is the best known. The Cape lobelias are, by contrast, diminutive but rather cute and, although there are 35 species in the Cape flora, only one has as yet made a significant contribution to horticulture. This is *L. erinus,* a dwarf species with naturally very variable flower

LOBELIAS. Our dear little friend, L. erinus, is the centre of the group, from which are derived the bright blue, deep indigo, grey, and rosy-flowering varieties in request everywhere for marginal lines and edgings; the very perfection of bedding plants, which any one may grow with but a trifling exhibition of skill and patience, with the aid of a glass structure of some kind or other, it scarcely matters how rough and simple.

Shirley Hibberd, *The Amateur's Flower Garden,* 1871.

Long-time denizens of border, window box and hanging basket, the profusion of lobelia cultivars, of which this is a typical if anonymous example, are descended from a single, highly variable Cape species, *L. erinus.*

Whenever a subject of floral interest presents itself to my notice, it affords me much pleasure to be able to offer a few remarks upon it; and my attention is now directed to the extremely beautiful though little cultivated genus Mesembryanthemum, having a bed of them growing in the open ground, under a south aspected wall, and which, during sunny days, compose a blaze of beauty. Besides which I grow 200 plants in my greenhouse. There are upwards of 300 distinct species and varieties grown in this country, and I possess 162 of them in my collections, all of which possess some peculiar claims to beauty and interest, both in foliage and flowers. Producing annually an immense number of flowers of the most brilliant colours, and yet of the most extensive variety, having thick, fleshy foliage, of a most singular and interesting character, and being besides most easily cultivated, this beautiful genus appears to me to possess charms and merits of a more than ordinary nature; and I am at a loss to imagine how many cultivators can willingly neglect, or wilfully despise a genus of plants which certainly deserves to rank amongst the most pleasing and delightful of natures productions.

Joseph Hartley, *The Floricultural Cabinet,* 1856.

Their abundance and the relative ease with which seeds and live plants could be collected and shipped ensured that mesembryanthemums or 'fig marigolds' were well represented in the gardens and greenhouses of the eighteenth and nineteenth centuries. More than 300 species were listed by Sweet in 1826, including many which had been in cultivation for over a century. This plate from Weinmann's *Phytanthoza Icongraphia* depicts a variety of typical species, including the yellow form of Hottentot fig *Carpobrotus edulis,* now a major invasive pest in temperate coastal areas elsewhere in the world where it has been introduced.

colour, ranging from dark blue through pink to white. This variability and its ease of propagation from seed have contributed to its success. *L. erinus* is often described rather prosaically as the 'edging lobelia', and was particularly popular during the days of the Victorian bedding regime, corralling the pelargoniums or placed with painful precision amongst other low-growers.

There are now five series within the *L. erinus* cultivars: 'Cascade', 'Moon', 'Palace', 'Regatta', and 'Riviera'. These are grouped according to their habit, which can be compact or trailing, and their flower colour which, through years of selection, ranges from white to red and all shades of blue and purple. The individual flowers can be of one, two, or three colours, and the blooms on a single plant may vary considerably.

The Pride of Table Mountain

Given than there are almost 250 orchid species in the Cape, it may seem surprising that so few are grown in cultivation today. Even at the peak of Cape plant popularity in the early 1800s only 13 species were to be found in England.

The simple explanation for this is that the majority of Cape orchids are rather small and often inconspicuous and the region supports only a few large, demonstrative species to rival the glamorous ones from the tropics. This is not to say that they are not attractive, as many of even the tiniest ones are exquisitely constructed and charismatic beyond their stature. Nevertheless, it is unlikely, in the eyes of the public, that they will ever rival the tropical tribes – *Cymbidium*, *Dendrobium*, *Phalaenopsis* and their like – for sheer spectacle and, as importantly, ease of cultivation and floral stamina.

A few species, however, are respectably large and flamboyant. The best known of these is *Disa uniflora*, the 'pride of Table Mountain'. The *uniflora* epithet is misleading as each stem typically produces more than one bloom and seems to have been the consequence of the single-flowered specimen from which the Swedish botanist Peter Berg described the species in 1767.

The origin of the name *Disa* is unclear, although one explanation is that it refers to the legendary Swedish queen of that name. When the king was looking for a wife he required that she had to come into his court neither naked nor clothed, neither riding nor driving. So Disa arrived, resplendent in nothing but a fishing-net, with one foot on a sledge and one leg over a donkey. Adopting this novel approach to courtship, she won her king's hand.

With such bizarre taste and imagination, the couple were clearly made for each other. From the floral point of view, the fine lattice of markings on a disa's upper petals would seem to have reminded Berg of the fetching fishnets of the queen. Clearly some botanical taxonomists (and monarchs) need to get out more.

Flowering in mid-summer when there is little else in bloom, and parading elegantly by pools and streams, the bright red *Disa uniflora* has become Table Mountain's floral icon. In the old days it was heavily exploited and, until the introduction of protective legislation, was in

real danger of becoming extinct through the depredations of flower pickers and orchid enthusiasts. There are now plenty in cultivation to satisfy demands so there is no excuse (not that there ever was) for filching them from Table Mountain or the few other spots where the species grows in the south-western Cape.

The first *Disa* cross, reported in 1891, took the form of a hybrid between *D. uniflora* and *D. racemosa*, a relatively hefty purply-pink denizen of seeps and marshes. This was named '*D. veitchii*' after the Veitch nursery at which it was bred. Soon after, a second hybrid was obtained by crossing '*D. veitchii*' with another Cape species, *D. tripetaloides*. Raised by William Watson at Kew, it was named 'Disa Premier' and described as 'a handsome plant with a vigorous constitution.' Over the next 20 years or so, another dozen hybrids were produced but progress was slow and, in fact, after 1922 it ground to a complete standstill until 1981, when a cross between *D. uniflora* and *D. cardinalis* was produced and named 'Kirstenbosch Pride'.

Since then, interest in disas has blossomed. Over 300 hybrids have been created in the last 20 years, many through the dedication of a few enthusiasts, notably Professor Sid Cywes, a renowned paediatric surgeon as well as orchid expert, at the Cape.

Red disas and their cultivars are now widely available from specialist nurseries. While unlikely ever to attain the dizzy levels of popularity enjoyed by their flamboyant tropical cousins, their beauty and interest remain appealing to those prepared to put that extra bit of effort into their cultivation.

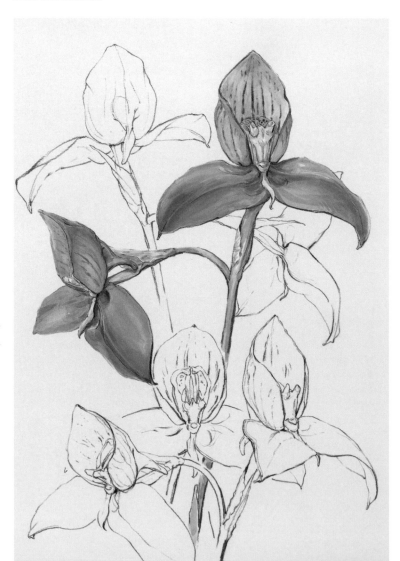

opposite

After a rather slow start, at least relative
to other Cape species, the cultivation and
hybridisation of *Disa* orchids has blossomed in recent years and there are now
over 300 varieties to choose from, including this *D. uniflora* 'Marguerite Kottler'.

right

Sometimes known as the phantom orchid,
Bonatea speciosa was first described as *Orchis
speciosa* by the younger Linnaeus in 1782.
Its present name was conferred by the
German botanist Carl Ludwig Willdenow
(1765–1812); *bonatea* commemorates
Guiseppi Antonia Bonato (1753–1836),
Professor of Botany at Padua. Found in
coastal scrub and forest edge from the
Cape to Zimbabwe, *Bonatea speciosa* can
grow to a metre high. In 1838 it was
included in a list of the 'finest flowering
species' of 'Terrestrial tropical orchideæ'
for English greenhouses.

CHAPTER THIRTEEN

The Smallest Kingdom

Of the thousands of plant species brought back from the Cape in the course of four centuries of botanical exploration, only a few hundred have found an enduring niche in our flowerbeds and greenhouses. Far more have fallen by the wayside after enjoying wide but fleeting commercial popularity or the short-lived attention of specialist plantsmen and collectors.

It has to be said that many Cape plants can be challenging to grow and the cosy domesticity of rich soils and fertilisers are no substitute for the demanding conditions in which they flourish on their home ground. That others will happily adapt to life in cultivation, given the necessary horticultural tweaking, is amply demonstrated by those described in these chapters. There is also no doubt that Masson, Bowie, Niven and their comrades have by no means exhausted the Cape's horticultural coffers and we can look forward to more outstanding contributions from a region which, for its size, supports the richest flora in the world.

Bounded to the south by the Indian Ocean, to the west by the cold Atlantic, and to the north and east by the almost desert-like expanse of the Karoo, the southern and south-western Cape represents an

ecological island, isolated and distinct from the rest of South Africa and the African continent. Within this island thousands of species of plants have evolved that are so unusual and so different from those found anywhere else in Africa and the rest of the world that the region has been afforded the accolade of a 'floral kingdom'.

Botanists have divided the world's vegetation into floral kingdoms according to the species or families of plants and their characteristics that are found growing naturally within particular areas. On this basis, there are six such kingdoms. With the exception of the Cape, these cover vast areas and may occupy entire countries, such as the Australian Floral Kingdom, or even greater landmasses such as the Boreal Kingdom that stretches across much of the northern hemisphere. When all its strips, patches and scattered fragments are combined, the Cape Floral Kingdom occupies approximately 88,000 km². Being far and away the most diminutive of the world's floral kingdoms, the Cape conveniently provides its own title for our book.

A picturesque mix of broad coastal plains, deep inland valleys and some spectacularly rugged mountains (the highest being 2,325 m), the Cape Floral Kingdom clings precariously to the edge of the African continent, occupying a narrow crescent of land 1,000 km long and not much more than 200 km at its widest.

Few parts of the world can approach the Cape Floral Kingdom for sheer botanical richness. At the last tally, 9,086 species had been recorded within this rocky ribbon of land. On a continental level, it occupies 0.5% of the land area of the Africa, yet it supports 20% of its plant species. At a smaller scale, the Cape Peninsula (470 km²) has 2,285 species, while Table Mountain alone supports 1,500 species in its 57 km². In comparison, about 1,400 species of plant occur naturally in the 308,000 km² of the British Isles.

More impressive, perhaps, even than this extraordinary richness is the fact that almost 6,200 (70%) of the Cape flora's species are endemic, in

opposite
Land of daisies and bulbs – the floral brilliance of Nieuwoudtville and the Hantam National Botanical Garden. White *Hesperantha rivulicola* and orange harlequins *Sparaxis tricolor* flower in profusion in seasonal seeps and rivulets.
John Manning/South African National Biodiversity Institute

To give some idea of the botanical riches of the country, I need only state that, in the short distance of one English mile, I collected in four hours and a half, one hundred and five distinct species of plants, even at this unfavourable season; and I believe that more than double that number may, by searching at different times, be found on the same ground.

William Burchell, *Travels in the Interior of Southern Africa*, 1822.

other words they occur naturally nowhere else in the world. This is one of the main factors contributing to the region's status as a floral kingdom. Many of its endemic species also have disconcertingly small distributions and having a natural global range not much larger than, say, a tennis court (as is the case with quite a number of species at the Cape), renders such plants particularly vulnerable to extinction through even small-scale human interference. The building of a single house (with or without tennis court) could, therefore, wipe out a species.

From all perspectives the Cape Floral Kingdom (or Cape Floristic Region as it is sometimes also known) is a remarkable and unique area – the world's 'hottest hotspot' of plant biodiversity and levels of endemism.

Plants in their Places

The Cape Floral Kingdom is made up of six major vegetation types, with fynbos the largest and supporting the most plant species. The others are renosterveld, strandveld, succulent Karoo, and kloof woodland and forest. While these are not as extensive or rich as fynbos, they all contribute to the floral diversity of the region and often support wildlife that might otherwise be absent or scarce.

Renosterveld, which may be considered almost a subspecies, as it were, of fynbos, is found on the fine-grained, fertile shale soils of the long, gently undulating valleys that separate the fynbos-clad and often spectacularly rugged and contorted Cape Fold Mountains that run parallel to the coast north and east from Cape Town. Its soils being too

left
Where it all began. To the west (left) of Cape Point and its lighthouse, commissioned in 1919, the Cape of Good Hope broods at the corner of the continent. This landmark, the south-western extremity of Africa, was officially named 'Cape of Good Hope' as recently as 1957, the appellation having previously referred geographically to the whole Cape Peninsula, including Table Mountain, and politically to the Cape Province.   *Peter Ryan*

top
The first barriers to travelling east from Cape Town are the Cape Folded Belt mountains that flank the False Bay coast, here running down to the sea at Kogelbaai.   *Peter Ryan*

above
Pink-flowered *Podalyria buxifolia* growing on the gentle summit plains of the fynbos-clad mountains at Baviaanskloof ('Baboons' Ravine') in the eastern Cape.   *Richard Cowling*

The top ten plant families and genera in the Cape flora

Plant family	Number of species	Endemic species	Plant genus	Number of species	Endemic species
Asteraceae (Daisy)	1036	655	Erica (Ericaceae)	658	635
Fabaceae (Pea)	760	627	Aspalathus (Fabaceae)	272	257
Iridaceae (Iris)	661	520	Pelargonium (Geraniaceae)	148	79
Aizoaceae (Mesembs)	660	525	Agathosma (Rutaceae)	143	138
Ericaceae (Heath)	658	635	Phylica (Rhamnaceae)	133	126
Scrophulariaceae (Figwort)	418	297	Lampranthus (Aizoaceae)	124	118
Proteaceae (Protea)	330	319	Oxalis (Oxalidaceae)	118	94
Restionaceae (Cape reed)	318	294	Moraea (Iridaceae)	115	79
Rutaceae (Citrus)	273	258	Cliffortia (Rosaceae)	114	104
Orchidaceae (Orchid)	227	138	Senecio (Asteraceae)	110	58

From *Cape Plants, A Conspectus of the Cape Flora of South Africa* by Peter Goldblatt and John Manning, 2000.

The region also contains four endemic species of mammal out of the 90 that are found here, and six endemic species of bird out of 320 that are resident or regularly occur. Sixteen of its 40 species of frog and toad, and 22 of its 100 or so reptiles (lizards and snakes) are endemic, as well as 14 out of 35 species of freshwater fish. About 70 of its 230+ butterfly species are endemic. The number of other endemic insects and invertebrates is unknown but likely to be in a proportion that rivals the plants.

nutritious for its own good, only tiny fragments, amounting to as little as 2% of its original area, have survived the transformation to agriculture in 350 years of European settlement. Renosterveld is particularly rich in bulbous plants and many of these, including a number of strikingly beautiful irises and enigmatic orchids, have become some of the most endangered species in the Cape Floral Kingdom.

Strandveld vegetation is a feature of the coast, occurring on limestone and on shell-sands that have been recently (geologically speaking) wind-blown to form dunes, or are the remains of raised beaches formed by fluctuating sea levels over the last few tens of thousands of years. Strandveld supports large numbers of evergreen succulent plants and typically fine displays of spring annuals such as daisies. A variety of evergreen shrubs, some attaining the size of small trees, is also found here, and as a whole it represents the southern exponent of the subtropical thicket that extends north up the east coast of South Africa. In the east of the Cape Floral Kingdom, this subtropical thicket is known as valley bushveld as it occurs in dense, shrubby stands along the river valleys.

Succulent Karoo may sound like a contradiction in terms, for a drier and dustier place than the Karoo would be hard to imagine. But it is to this very drouth that its many plant species are adapted. Growth here is restricted to the cooler, damper months of winter and spring and any moisture that the plants can store up for a non-rainy day is held in reservoirs in the juicy, fleshy stems or leaves, hence the 'succulent' label. Small outliers of this vegetation occur in the north and east of the Cape Floral Kingdom, with larger expanses in the rain-shadowed, summer-parched mountain valleys and basins along its northern length.

Woodland is a rather restricted habitat in the region. Being drought- and fire-prone the local environment does not suit trees, and a perennial water supply and protection from fire are necessary for those that do occur here. The Cape climate has dried out over the past few million years and trees have consequently become confined largely to the east of the region. Where they do persist in the west, trees are mainly restricted to scree slopes and kloofs. Further east, in contrast, where rainfall occurs year-round, there are some impressive tracts of indigenous trees, the most famous of which is the Knysna Forest with its venerable yellow-wood trees and relict population of elephants.

Fynbos is the most extensive vegetation type, covering just over 41,000 km², or about half the area of the Cape Floral Kingdom. It is also the most important in terms of richness, contributing perhaps as much as 80% of the species total to the area as a whole. Given this prominence, fynbos has, understandably but inaccurately (and to the annoyance of botanical purists), become synonymous with the entire suite of varied but distinctive and unique plant communities that make up the Cape Floral Kingdom.

Fynbos is the characteristic vegetation of the Cape Floral Kingdom's rocky hills and slopes. It grows in extremely coarse, acidic, and nutrient-poor quartzitic sandstone soils derived from the slow erosion of the Cape Fold Mountains. The soils also contain very little organic matter, and the vegetation is naturally fire-prone and liable to go up in flames every few years.

Considerable confusion has arisen over the years regarding the distinctions between fynbos, the Cape Floral Kingdom and the fynbos biome. Fynbos ... is a type of vegetation uniquely characterised by restioids, and confined to the nutrient-poor soils of the southwestern and southern Cape. The Cape Floral Kingdom is a floral province defined on the basis of the number of plant species which grow there and nowhere else, i.e. are endemic to the area ... The highest category of plant community recognised in the world, a biome is defined in terms of climate and the dominant growth forms in the vegetation ... The fynbos biome is defined by moderate to high amounts of winter rain and a predominance of low to medium-height shrubs, and as such includes three vegetation types: fynbos, renosterveld and subtropical thicket.

Richard Cowling and Dave Richardson, *Fynbos – South Africa's Unique Floral Kingdom*, 1995.

top
Red and yellow conebushes (*Leucadendron* species), tall restios, and a variety of small-leaved shrubs including white-flowered Blombos *Metalasia muricata* dominate this fynbos scene near Worcester.

above
A track leads through a valley in the Cederberg, a spectacular and botanically-rich wilderness area some 200 km north of Cape Town, that is popular with climbers and hikers.

Essentially a heathy shrubland, fynbos is distinguished by a general scarcity of grasses and annual plants. It is also characterised by four botanical elements or growth forms. These are the proteoid (protea or protea-like broad-leaved shrubs); ericoid (small shrubs such as erica heaths with narrow, often in-curled leaves to minimise water loss); restioid (wiry, reed-like, leafless plants) and geophytic (bulb-like plants) elements. Together these form the most distinctive botanical and, to a large extent, visual features of the fynbos landscape. The restioid element is taken to be the definitive characteristic, without which no vegetation could claim to be fynbos.

Along the length of the Cape Floral Kingdom the climate ranges from a year-round, or mainly summer rainfall regime in the east, to a winter rainfall one in the far west. In the western part of the Cape Floral Kingdom a Mediterranean-type climate prevails with warm, dry summers and cool, wet winters. Winter and, particularly, spring are a joy here, with calm, warm days interspersed with wild winds and invigorating rains that stimulate the plants to sprout and bloom. Summer is warm, sometimes hot, and the Cape Doctor, the raging south-easterly summer wind that clips the coastal scrub and blows blustery Capetonians up and down their city streets, can be a bit wearisome. Down at Cape Point, on the tip of the Peninsula, the wind is even worse, registering gale force from one direction or another on 250 days in an average year. It's no surprise that Dias may have christened it the Cape of Storms.

Mediterranean-type climates prevail not only in those areas of southern Europe, the Middle East and North Africa that bound the Mediterranean Sea, but also in parts of the world at similar latitude, be they north or south of the equator, such as California and regions of

Two strong winds chiefly prevail on this southernmost promontory of Africa. The one blows boisterously almost every day in summer, which is called the Good Season (Goede Mousson); the other in winter, which is called the Bad Season (Quaade Mousson). The south-east wind is violent and attended with dry and very fine weather; the north-west is tempestuous, and for the most part, accompanied with showers of rain. The former brings short and violent gales, following close upon each other, which often increase to that degree of force, as to blow up not only dust and sand, but also gravel and small pebbles in to the face of such as are exposed to it, who, being neither able to see or go forward, must either stand still, or else throw themselves down on the ground. On such occasions strangers frequently exhibit ridiculous scenes, their hats, wigs, or hair bags being carried away by the wind the whole length of the streets.

Carl Peter Thunberg, *Travels at the Cape of Good Hope 1772–1775*.

above
The diagnostic plants of fynbos, and the quintessential species of the Cape Floral Kingdom, are proteas, ericas, restios and bulbs. It would be hard to choose a typical protea as the family is so diverse in shape and form. *Protea pityphylla* ('pine-leaved') is at least becoming more familiar to gardeners as it takes well to cultivation. The species was discovered in July 1883 by Alfred Bodkin (1847–1930), a mathematics teacher and dedicated amateur botanist, and is restricted to a small mountainous area near Ceres.

opposite
Restio dispar, a member of the Restionaceae (Cape reeds). If the vegetation does not include a representative of this family, then it cannot claim to be fynbos!

Chile and south-western Australia. These regions and the Cape share many ecological features, of which the most important is fire.

Fire and Phoenixphytes

Fire is the major ecological driving force in the Cape Floral Kingdom as a whole, and one to which the plants have evolved some remarkable survival strategies. It is most prevalent in fynbos, but under the right conditions any of the vegetation, with the exception of succulent Karoo, will burn if put to the torch, naturally or otherwise. Under natural conditions, the main sources of ignition are lightning-strikes and sparks from rockfalls following earthquakes or rolling stones dislodged by animals such as baboons. Nowadays the main spark, and not a very bright one, is barbecue- or pyromanic-man.

While it might appear counter-intuitive to imagine that fire is actually good for plants, it is recognised that not only are fynbos plants adapted to surviving fire, they are dependant upon it to complete their lifecycles and, therefore, rely on it for their very existence. An extensive suite of fynbos plants is called 'Phoenixiphytes' because they rise from their own ashes, the latter providing recycled nutrients in an environment that is markedly short of them.

Fynbos plants have evolved a number of strategies by which to survive fire: geophytes that regenerate from underground storage organs (such as bulbs and tubers); resprouters that regrow from woody rootstocks; thick bark that protects dormant buds deep in the stem; and shrubs that are killed by fire but regenerate from seed stored in protective seedheads, such as cones, which open and release the seed after the fire. This latter phenomenon is known as serotiny.

below
Cobra lily or kapelpypie *Chasmanthe aethiopica* was introduced into Europe in 1635. It has become a popular garden subject, but less so than its close relative and ubiquitous horticultural favourite *Crocosmia* ('Montbretia'), seven species of which occur in South Africa, but none naturally in the Cape Floral Kingdom.

above
Chasmanthe aethiopica growing just above the beach on the southern Cape coast. *Richard Cowling*

Fires can occur at any time of the year, particularly in the east of the region, but the majority are in autumn when the prolonged heat and drought of summer have left the vegetation and leaf-litter tinder dry. The major source of ignition for the last 125,000 years or so is Man, although lightning strikes and rockfalls are important natural causes. *Peter Ryan*

About 3,000 species, some 30% of the Cape flora, are 'myrmecochorous', meaning 'seed-dispersal by ants'. Seeds of these plants are not stored in a seedhead, but ripen and fall within two months or so of pollination. The seeds have a fat, fleshy seed-coat or appendage, called an elaiosome, that is very attractive to ants. When the seeds drop to the ground they are carried off by foraging ants to their underground nests, where the elaiosome is eaten but the seed itself is unharmed. Deep in the nest the seeds are protected from predators and the ants also secrete powerful fungicides to prevent stored seed from rotting. Germination within the nests only takes place after the above-ground vegetation has been burnt.

A variety of fire-related factors, including temperature, increased light, and the release of nutrients, has been invoked to explain the breaking of dormancy and subsequent synchronised germination of myrmecochorous and other seeds in post-fire fynbos. The most important cue has now been identified as smoke, with a chemical called butenolide being the critical constituent.

Myrmecochory and serotiny both serve to protect nutrient-rich seeds from fire and from being eaten by birds or rodents in the otherwise notoriously nutrient-poor environment. These adaptations also demonstrate that fire is, at appropriate intervals, a natural process in fynbos. Furthermore, fire and disparate nutrient availability have, over millions of years, been major evolutionary forces that, when set in a highly dissected landscape, have shaped and driven the plants to extraordinary levels of speciation, the hyperdiversification found in the Cape Floral Kingdom today.

Scorched earth and charred stems do not give the impression that fire in fynbos is anything but a bad thing. Note, however, the russet drifts of protea seeds released from cones that were split open by the heat. These soon germinate, and with no competition from older plants for water and sunlight and with a flush of nutrients from the ash, the seedlings will flourish.
Brian van Wilgen

The underground bulbs of plants such as these *Cyrtanthus* (top), *Haemanthus* (left), and *Brunsvigia* (above), respond almost instantaneously to the passage of a fire. In the Jonkershoek Valley near Stellenbosch, *Cyrtanthus* had not been seen in bloom since the last fire swept through the fynbos 40 years ago. When the valley was burnt again after all this time, many *Cyrtanthus* and other 'phoenixiphytes' flowered within a few days, then quickly set seed and died back to await the next conflagration years or decades hence.
Brian van Wilgen

The bursting into bloom of thousands of *Watsonia borbonica* turns a recently burnt hillside into a forest of pink at Moordenarskop ('Murderer's Peak') in the Hottentots' Holland. Synchronous flowering after a fire attracts large numbers of nectar-feeding insects and birds, maximizing the chances of each flower being successfully pollinated.

So much seed is then produced that there may be more than any mice and birds can consume. The prospects of individual seeds surviving to germinate are, therefore, improved by sheer weight of numbers. This ecological phenomenon is called 'predator satiation'.
William R. Liltved

Red-hot pokers *Kniphofia* bloom *en masse* in the mountains following a fire the previous autumn.
William R. Liltved

Important pollinators of many Cape plants, especially proteas and ericas, are the Cape sugarbird, orange-breasted sunbird, and southern double-collared sunbird. The first two are endemic to the region. *Peter Ryan*

The enormous variation in shape, colour and size of Cape flowers has been interpreted as the plants' evolutionary response to securing a pollinator amongst fierce competition from other flowers. Orchids such as *Bonatea speciosa* are pollinated by hawkmoths (note the yellow pollen sacs adhering to its 'chin'), while a long-tongued *Moegistorhynchus* fly visits a pelargonium. *Steve Johnson*

The African porcupine is the world's largest rodent. Evidence that one has taken an interest in a particular patch of veld or has visited your garden is usually confined to a scattering of quills and a complete demolition job of any plant that is remotely edible, although bulbs and tubers are preferred. The Cape Floral Kingdom is rich in reptiles and amphibians. The angulate tortoise (this is a youngster) is the most common and widespread of the six tortoise species found here. The western leopard toad is also known as the August frog or snoring toad on account of, respectively, its breeding season and vocalisations.

Jan 31. I saw today a curious little bird, which seems to be the Cape Honeysucker [Sugarbird] ... of a brownish colour, with a tail of many fine, long, soft, wavy feathers, three or four times as long as the body; the beak slender and arched. They do not ... feed on the wing like humming birds, but while perching; their flight is short, jerking, and unsteady, as if they were embarrassed by the length of their trains.

Charles Bunbury, *Journal of a residence at the Cape of Good Hope; with excursions into the interior, and notes on the natural history, and the native tribes*, 1848.

Visitors today, in addition to enjoying its wonderful flowers and other wildlife, can see for themselves that the Cape Floral Kingdom has been extensively changed by human actions. Agriculture and urban and industrial development have transformed large areas, and invasion by introduced plants threatens those parts that have not disappeared under concrete or the plough.

The Trouble with Aliens

Gardeners have done a huge amount to beautify the artificial environments that they have created for themselves and others to enjoy. It is only fair to say, however, that they have done at least as much, and probably a great deal more, to blight the natural environment.

By this I don't mean the unsustainable exploitation of stocks of wild plants, although that has also been significant over the years. More importantly, they have spread their favourite flowers, shrubs and trees to parts of the world where they do not naturally occur. While many of these uprooted expats have sat contentedly in their bed, border or pot ever since, others have been decidedly more adventurous. Having hopped over the garden wall or crept through the picket fence they have gone on to run riot in their country of adoption, swamping the native plants as they careened over the countryside. Few places in the world have been subjected to this destructive tidal wave as ruthlessly as the Cape Floral Kingdom.

'Send us anything that will grow', implored Jan van Riebeeck of his masters back in Holland following his arrival at the Cape in 1652. The imports began with the cereal crops, fruits and vegetables required to sustain the resident population and to supply the passing ships. Familiar flowers were also imported to brighten up the gardens of the homesick settlers.

It was not until the early years of the nineteenth century, however, that the bulk of plants that were to become major invaders were introduced. This was the period when the British colonisation of Australia was in full swing and there was an enthusiastic interchange of products and commodities between the countries including numerous shrubs and trees introduced to the Cape for ornamental or practical purposes.

Over 100 species of alien (non-native) plants have now become invasive in the Cape Floral Kingdom and, while this is not a high number relative to the richness of native species, these aliens have overwhelmed the Cape with an alacrity and potency that beggar belief.

Australian wattles make up almost half the list of the most successful, that is to say damaging, invasive plants here. Rooikrans *Acacia cyclops* is now the most widespread invasive wattle and, being a species of sandy flats and dunes in its homeland, it has been particularly successful in invading this type of habitat along the Cape coast.

Other aliens are no less efficient in their subjugation. Silky hakea *Hakea sericea*, introduced from New South Wales as a hedging plant to the Cape Peninsula in the 1830s, took just over 100 years to infest almost 5,000 km² of mountain fynbos.

Comparatively few of the gems which strew the veld are grown in our gardens, and unless we make some provisions for the future it is not at all improbable that any one, ten years hence, wishing to see Cape bulbs, will have to seek them at Kirstenbosch or Kew or in the nurseries gardens of Holland – so rapidly are many varieties vanishing under the advance of civilisation.

Dorothea Fairbridge, *Gardens of South Africa*, 1924.

Many Australian wattles, such as long-leaved wattle *Acacia longifolia*, have been introduced to the Cape over the years and found it very much to their liking. They might look pretty, but they are rapidly overwhelming the native species of the Cape Floral Kingdom. Indeed, invasive alien plants such as these represent the single most important threat to the region's unique flora.

Plants are not the only inhabitants of the Cape Floral Kingdom that are endangered. The majority of large mammals in the region, including black rhinoceros, elephant and hippopotamus, were wiped out within a few years of European colonisation. Narrowly escaping complete extinction was the Cape mountain zebra. By the 1930s, hunting and habitat destruction had reduced the population to fewer than 50 and governmental support was sought to save them. The Minister of Lands, General Jan Kemp, declined, declaring that they were just 'a lot of donkeys in football jerseys'. Other people were more sympathetic, and with careful management and protection the population now stands at over 500.

It may be naturally supposed that, in a country abounding with the most beautiful flowers and plants, the gardens of the inhabitants contain a great number of its choicest productions; but such is the perverse nature of man's judgement, that whatever is distant, scarce, and difficult to be obtained, is always preferred to that which is within his reach, and is abundant, or may be procured with ease, however beautiful it may be. The common garden flowers in Europe are here highly valued; and those who wished to show me their taste in horti-culture, felt a pride in exhibiting carnations, hollyhocks, balsamines, tulips, and hyacinths, while they viewed the elegant productions of their hills as mere weeds.

William Burchell, *Travels in the Interior of Southern Africa*, 1822.

The rhinoceros-bush (a species of stoebe) a dry shrub, which is otherwise used to thrive on barren tracts of land, now begins to encroach more and more on such places as have been thoroughly cleared and culti-vated. When I asked the country people the reason of this, they would lay the blame on their sins. Their consciences, probably, informed them, that there was great reason for so doing. One of their sins which most merited this punishment, as having con-tributed most to evil, might, in this case, be reckoned their want of knowing how to dress properly the soil they occupied, and to manage it to best advantage.

Anders Sparrman, *A Voyage to the Cape of Good Hope* 1772–76.

'The emigrant plant tends to leave its troubles back home'. Immigrant plants leave their natural enemies (in particular herbivorous insects) at home and enjoy a relatively stress-free existence in fynbos, using resources normally used in repairing damage inflicted by herbivores and pathogens for growth, flowering and seed production. Alien plants can therefore grow faster and produce more seeds at a younger age than indigenous fynbos plants, giving them a crucial competitive edge over their indigenous cousins.

Steven Higgins, *Veld & Flora*, 1995.

The success of the invasive aliens can be ascribed to a number of factors. Perhaps most obvious is the fact that they originate in countries that enjoy the same environmental conditions, notably the Mediterranean-type climate, as the Cape. They are, therefore, already adapted to life under such conditions, but without all the pests and diseases that afflict them back home, so the perfect ecological conditions exist for invading their new territory. Such has been their impact that more than half the endangered plant species in the Cape Floral Kingdom today can count alien invasion as the primary threat to their existence.

Cape Plants Abroad

A corner of Kirstenbosch National Botanical Garden displays a variety of South African plants, many of them from the Cape Floral Kingdom, that have become troublesome weeds elsewhere. Ironically, but not perhaps surprisingly, this includes some of the most attractive and popular garden plants that the country has given the world. So what is a beautiful wildflower garden in Cape Town is a conservation nightmare somewhere else. Beds of *Agapanthus*, for example, remind us how these plants have become invasive in Australia and New Zealand, the Channel Islands, and Madeira.

Australia seems to have borne the brunt of these undesirable, if not invariably accidental, tourists. The splendour of Kirstenbosch notwithstanding, it is wryly said that if you want to see fynbos plants at their best, go to Australia. Introduced as ornamentals and escaping into the wider countryside, not least through the dumping of garden rubbish, bulbs have done particularly well with arum lily, gladioli, babianas, freesias, chincherinchees and sparaxis now highly invasive and ecologically damaging weeds there.

By the same token, a wide range of popular introduced ornamentals, lovingly tended in gardens at the Cape, are beginning to indicate that they're not content with languishing in captivity. Plants such as pampas grass, Spanish broom, purple loosestrife, and New Zealand christmas tree, amongst many others, if not quickly eradicated inside and outside gardens, will almost certainly become a major threat to native plants and other wildlife as they romp through the Cape Floral Kingdom. This problem will almost certainly be compounded by global warming which will enable more alien species to cross the ecological threshold and become invaders.

It is up to nurserymen and gardeners to take a responsible attitude and not to import nor cultivate plants that might carry even the slightest risk of becoming invasive, notably species from similar climate-zones of the world. Most of tomorrow's invasives are in today's gardens – Cape flora beware.

Discovered on the west coast in 1971 by Johan Loubser, *Moraea loubseri* (right) is confined to a solitary granite outcrop, much of which has been removed by quarrying. Fortunately, it grows well in cultivation. *M. polystacha* (left), in contrast, occurs widely in southern Africa and is also a rewarding pot-plant. Although the individual blooms are short-lived, the flowering season can last for eight weeks. Many horticulturally-challenging and less charismatic plants will not benefit from 'conservation through cultivation' and will fade quietly into oblivion as humans and invasive aliens continue to take their toll on the Cape Floral Kingdom.

opposite
Thomas Stokoe (1868–1959) with *Mimetes stokoei* in a billycan. The 'extinction' and rediscovery of this enigmatic species is described in *T P Stokoe, the man, the myths, the flowers*, a wonderful book that captures the magic of his plants and the Cape mountains that he loved.
Bolus Herbarium/Colin Paterson-Jones

opposite right
In 1954, Elsie Esterhuysen found an unknown protea in the Langeberg. Material to identify it was eventually collected in 1983 by Esterhuysen, here with John Rourke (centre) and John Winter (right, directly behind the new species). Rourke created a new genus *Vexatorella* ('little troublemaker') for it because it 'upset the previously stable classification' of the Proteaceae.
Colin Paterson-Jones

CHAPTER FOURTEEN

Back to its Roots

By the late nineteenth century, the collecting of Cape plants for cultivation had diminished. Not only had the novelty worn off, but expeditions to highland areas of the Far East, in particular, had revealed a cornucopia of exotic species that were soon to become staples of the herbaceous border and shrubbery. An important factor that heightened the popularity of these new arrivals was that their countries of origin experienced a similar climate to that of western Europe and North America, the two largest gardening markets. It being a great deal easier to grow a peony from China than a protea from the Cape it was not surprising that, with few exceptions, Cape plants became largely neglected.

This is not to say that interest in the Cape flora had vanished altogether; far from it. The vacuum of curiosity created by the decline in horticultural exploitation was gradually filled by the scientific investigation of the Cape flora and the emergence of predominantly resident botanists.

The Botanical Nonagenarians

While botanical science at the Cape was in the ascendancy, it was supported by traditional collecting that, while less intensive than that of the past was, in many respects, more demanding. Areas were targeted that had not been visited much, or at all, by the historical plant hunters, and therefore required more effort to identify and get to since many were far from the by now well-beaten tracks. To make a useful contribution to botany, it was no longer a case of casually picking a protea at the roadside.

Two of the outstanding collectors of the tewntieth century were Thomas Stokoe and Elsie Esterhuysen. They were not only plant hunters in the finest traditions, but set new standards of their own through extraordinary hard work and sheer love of their subject.

Thomas Stokoe emigrated to Cape Town from Yorkshire in 1911. He combined mountaineering with a passion for fynbos and spent 48 years climbing and botanising amongst some of the remotest peaks in the Cape. He collected over 16,000 plant specimens (although it is noted that 'he was not methodical in his collecting and rarely numbered his plants'!), including many new species. Thirty of these were named after him, including a number of ericas and proteas. Described by Conrad Lighton, author of *Cape Floral Kingdom*, as 'the world's toughest and sprightliest nonagenarian' he died in 1959 shortly after climbing in the Hottentots Holland to celebrate his 91st birthday.

Plant collecting for the horticultural trade, once indulged in with carefree and apparently limitless abandon at the Cape, is now strictly controlled. Two of the most respected and knowledgeable collectors are Rod and Rachel Saunders of Silverhill Seeds. Here they admire the chincherinchees *Ornithogalum thyrsoides* on Devil's Peak above Cape Town.

Andrew Harvie

On an expedition in December 1928, Thomas Stokoe and his companions found no fewer than eight new plant species, including *Protea aristata*. This stocky bush grows on rocky sandstone slopes in the mountains near Ladismith, which has adopted the flower as its town emblem. As a cut-flower it has good looks but unfortunately smells terrible as the leaves emit a hydrogen sulphur-like pong.

Game guard Willie Julies found this spectacular new protea, *Mimetes chrysanthus*, in a remote corner of the Gamka Mountain Nature Reserve in 1987, proof that not all species awaiting discovery are obscure and undemonstrative.

Colin Paterson-Jones

Elsie Esterhuysen also demonstrated that hills and plants can be good for the constitution; she exceeded Stokoe in both longevity (she died in 2006 at the age of 94) and plant collecting but, like him, concentrated on the plants of the mountains. Very frugal by nature, she would happily spend days at a time in the field sustained only by raw carrots and cabbage, with the occasional indulgence of weak tea and powdered milk. Described by John Rourke as no less than 'the most outstanding collector ever of South African flora', she collected over 36,000 specimens and discovered more than 100 new species, including some 25 ericas, and is commemorated in the names of many of them.

Into and during the twentieth century, therefore, the Cape could boast not only a hugely impressive list of plant species but a continuing, and surprisingly high rate of new discoveries through the efforts of Stokoe, Esterhuysen and others. In the 1970s, however, there was an almost embarrassed realisation that, despite the fact that a globally important and threatened ecosystem was perched precariously here on the edge of Africa, there was little understanding of its fundamental ecology – what made it tick.

Fynbos Discovery

The establishment, in 1977, of the Fynbos Biome Project sought to pull together the threads of previous and current research that, while useful and insightful in its own way, was disparate and unsystematic. The project thus aimed to optimise research efforts through co-ordination, to synthesise existing information on fynbos, identify gaps, and focus efforts on providing 'sound scientific knowledge of the structure and functioning of constituent ecosystems as a basis for the conservation and management of … fynbos.' While this did not cover the whole Cape Floral Kingdom, by confining itself to fynbos it automatically addressed the vegetation type with the greatest area and species richness.

The Fynbos Biome Project ran for ten years, providing a springboard for the wealth of research currently being undertaken in the region by a number of institutions and a multitude of scientists. As well as being useful academic investigations in their own right, these studies provide fascinating insights into how the Cape flora and its animal cohabitants have adapted and how they interact and co-exist. Most importantly, they provide the information necessary to conserve it and its integral parts and processes most effectively.

Disciplined research has also made it possible to communicate the wonder of the Cape Floral Kingdom to a wide audience. Growing local and overseas audiences were won not just through scientific publications, accessible to and, often enough, understood by only relatively few people of academic inclination, but through increasing numbers of popular publications in the form of books and articles, notably in the Botanical Society of South Africa's magazine *Veld & Flora*.

An important milestone was the publication in 1984 of what we

Bill Liltved follows in the finest tradition of the dedicated amateur botanist (with which the Cape is well endowed). He has an all-consuming interest in Cape orchids and in his search for new and rare species he has covered thousands of kilometres, much of it on foot, to parts of the region that have been rarely, if ever, visited by plant hunters. His book *The Cape Orchids* is the spectacular and monumental result of more than 20 years of dogged orchidology. Here Bill admires cluster disa *Disa ferruginea* at Silvermine on the Cape Peninsula.
Jim Holmes

Kirstenbosch garden foreman Adonis Adonis with *Erica verticillata*. This species was driven to extinction by the ever-expanding Cape Town in the mid-twentieth century. In the 1980s, however, a plant was discovered in Pretoria, most likely introduced there in the 1940s. Plants were then found growing at Kew and in the Belveidere Gardens in Vienna. The former were probably descended from seed sent by Masson, the latter by Boos and Scholl via the Schönbrunn Palace garden, over 200 years ago. Adonis discovered a single *E. verticillata* just beyond the Kirstenbosch boundary, the last surviving descendant of a small number planted in the gardens many years previously. Cuttings from all these sources have now secured the future of the species and allowed it to be reintroduced into the wild.
Anthony Hitchcock/South African National Biodiversity Institute

The loss of unique Cape plants to development, agriculture, and alien infestation far outweighs the discovery of new species. Occasionally, however, there are good-news stories that boost morale. This small protea, *Serruria foeniculacea*, was a presumed victim of urban development and aliens in its tiny range on the edge of the Cape Flats. A 14-year hunt by Howard Langley culminated in 1989 with his discovery of two plants growing in an area earmarked for development. These were transplanted into the nearby Rondevlei Nature Reserve and cuttings were successfully propagated at Kirstenbosch. *S. foeniculacea* remains critically rare in the wild and dependent upon careful protection and management at Rondevlei, virtually the last remnant of natural vegetation on the Cape Flats.

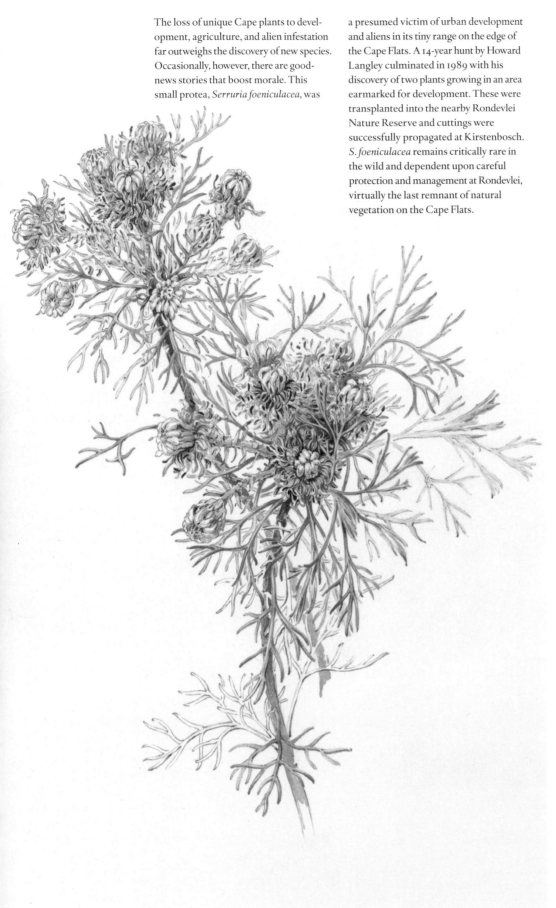

above

The Dutchman's Wallet

Smoke from fynbos fires contains a chemical called butenolide. This has been found to be the critical agent in breaking seed dormancy. By bubbling smoke through water, butenolide and other chemicals are dissolved. When paper discs soaked in the 'smoke water' are dried out and then added to water again they release the chemicals. Seeds watered with this solution are stimulated to germinate.

This technique has successfully triggered germination in seeds more than 200 years old. A Dutch merchant, Jan van Teerlink, acquired the seeds at the Cape in 1803, but the wallet containing them was confiscated when the British captured his ship. In 2005, the wallet was rediscovered in the National Archives by historian Roelof van Gelder. The seeds were identified by Matt Daws of Kew's Millennium Seedbank as coming from almost 40 fynbos species. A few that he treated with 'smoke water' germinated and grew into healthy shrubs providing a wonderful, living link with the early days of plant collecting at the Cape.
Wolfgang Stuppy

affectionately know as 'The Brown Book', an unassuming-looking volume entitled *Plants of the Cape Flora - A Descriptive Catalogue*. This lists every plant species known to occur in the region, with a brief description of the growth form and flower, their distribution and flowering time. Compiled by Pauline Bond and Peter Goldblatt, this is the 'Wisden' of the Cape Flora and, to plant geeks in South Africa and vicarious botanists overseas, simply indispensable. A new version, revised and updated by Goldblatt and John Manning, was published in 2000 under the title *Cape Plants – A Conspectus of the Cape Flora of South Africa*.

Another important volume, *The Ecology of Fynbos – Nutrients, Fire and Diversity,* was published in 1992. This synthesised and summarised the core components of fynbos, its diversity and endemism, reproductive ecology, plant-animal relationships, human settlement, non-native invasives, and ecosystem management. Each subject was written by experts under the overall editorship of Richard Cowling.

The pre-eminent fynbos ecologist of his generation, Cowling has contributed enormously to its understanding and appreciation. He is one of an increasing number of exceptionally talented and dedicated scientists and communicators who, over the years, have furthered our understanding of the functioning and requirements of fynbos and the Cape Floral Kingdom and have eloquently delivered their findings to academics and laymen alike.

While the ecologists are getting to grips with fynbos functioning, a number of dedicated botanists continue in the finest tradition of searching for and describing new species and sorting out the taxonomy of existing ones. If there was ever a danger that we would soon simply run out of new plants, it is worth noting that since the publication of the last volume of *Flora Capensis* in 1933, about 200 new species of *Erica* alone have been discovered in the Cape Floral Kingdom. In the first few years of the twenty-first century, a further 80 species, including 24 ericas and almost 30 members of the iris family, have been added to the Cape list.

Linnaeus once declared: *'Inexhaustum credo Cap. B spei esse plantarum speciebus; certe nulla Flora ditior erit'* (I believe the Cape of Good Hope is by no means exhausted of plant species; surely no other flora could be richer?). His words are, remarkably, as true today as they were when he wrote them in 1772. It's exciting and sobering to think that there are still new discoveries to be made in a part of the world that has seen such a long and distinguished succession of plant hunters scour its peaks, slopes, valleys and coastal plains.

Orange and white Wild Dagga *Leonotis leonourus* in flower at Kirstenbosch. National botanical gardens traditionally showcase plants from as many parts of the world as possible. With the exception of a few vintage oak trees, Kirstenbosch is unique in growing only South African plants.
Anthony Hitchcock/South African National Biodiversity Institute

And if in the vicinity of Cape Town, a well-ordered botanic garden of sufficient extent, were established, for the purpose of receiving plans which might casually, or even expressly, be collected in the more distant parts of the colony, the sum of money required for maintaining it would be but trifling, in comparison with the advantages which science, and the public botanic gardens of England, would derive from it.

William Burchell, *Travels in the Interior of Southern Africa*, 1822.

Visitors enjoy a colourful display of spring annuals at Kirstenbosch.
Liesl van der Walt/South African National Biodiversity Institute

Over the years it has become increasingly evident that the Cape Floral Kingdom is something very special. It is not just a quaint local phenomenon with a few pretty flowers, nor a convenient provider of pot plants. It is a globally unique and seriously threatened ecosystem for which we are all responsible.

In 2009 the Royal Botanic Gardens, Kew, from where the first plant collector, Francis Masson, was sent to the Cape, celebrated their 250th Anniversary. Also in that year, that same collector's venerable cycad, almost as old as Kew itself, was carefully repotted. These two events, so different and yet so intimately linked, remind us of the long and illustrious history that connects the Cape and Kew, together with anyone, anywhere, who has a Cape plant in their garden or greenhouse.

So, as you diligently tend your favourite pelargonium, savour the exquisite scent of freesias, or revel in the colourful exuberance of nemesias in full bloom, take a moment to remember those pioneering plant-collectors who brought them from a distant ancestral home, one that has given us so much over the years – the unique and beautiful smallest kingdom.

Perhaps we cannot do better than to enter this world of beauty through the gates of such a place as Kirstenbosch … There is a warmth in the air and a vital quality that quickens interest. The plants here are all South Africa's own, brought from the mountains and woods and fields.

Sarah Coombes, *South African Plants for American Gardeners*, 1936.

below left
The word's oldest pot plant, a cycad *Encephalartos altensteinii*, is repotted at Kew in 2009, almost 240 years since it was collected at the Cape by the gardens' first international plant hunter, Francis Masson.

below
Botanists from Kirstenbosch make their way home at the end of a successful day gathering plants and seeds in the western Cape mountains. More than 400 years after the 'thistle from Madagascar' was collected at the Cape, the task of studying and conserving its unique and beautiful flora goes on.
South African National Biodiversity Institute

Mindful of the role of the people of South Africa as custodians of the world's richest floral heritage, it is our mission to win the hearts and minds and material support of individuals and organisations wherever they may be for the conservation, cultivation, study and wise use of the indigenous flora and vegetation of southern Africa.

Mission statement of the Botanical Society of South Africa.

The establishment of Kirstenbosch National Botanical Garden in 1903 was underpinned by the simultaneous founding of the Botanical Society of South Africa, a condition of government funding being that the gardens also sought support from the public sector.

In this the Society has been singularly successful. It now has 16 branches in South Africa and over 15,000 members, communicating with these mainly through its flagship magazine, *Veld & Flora*. With articles by every level of botanist, from keen amateur to the most academic of professionals, and illustrated with high quality artwork and photographs, *Veld & Flora* succeeds in promoting the 'protection and conservation of our floral heritage'.

As well as its magazine, BotSoc (the Society's affectionate diminutive) also publishes regional wildflower guides, 10 of which cover parts of the Cape Floral Kingdom and without the appropriate one of which no one should venture into the veld. As important as its publications are BotSoc's environmental education programmes, reaching out to local communities, particularly in socially deprived areas, and to schoolchildren.

BotSoc's conservation work has expanded and developed over the years from, in its early days, addressing illegal flower picking, through initiating and co-ordinating alien clearance programmes, resisting potentially damaging construction and agricultural proposals, liaising with landowners, and developing strategic conservation programmes. A major initiative is the creation of its Cape Conservation Unit to facilitate conservation action in the two botanical hotspots of the Cape Floristic Region and the Succulent Karoo.

Apart from actually visiting the country, there is no better way for everyone overseas to learn about and enjoy South Africa's flora (especially the Cape Floral Kingdom) and to support its conservation than by becoming a member of BotSoc.

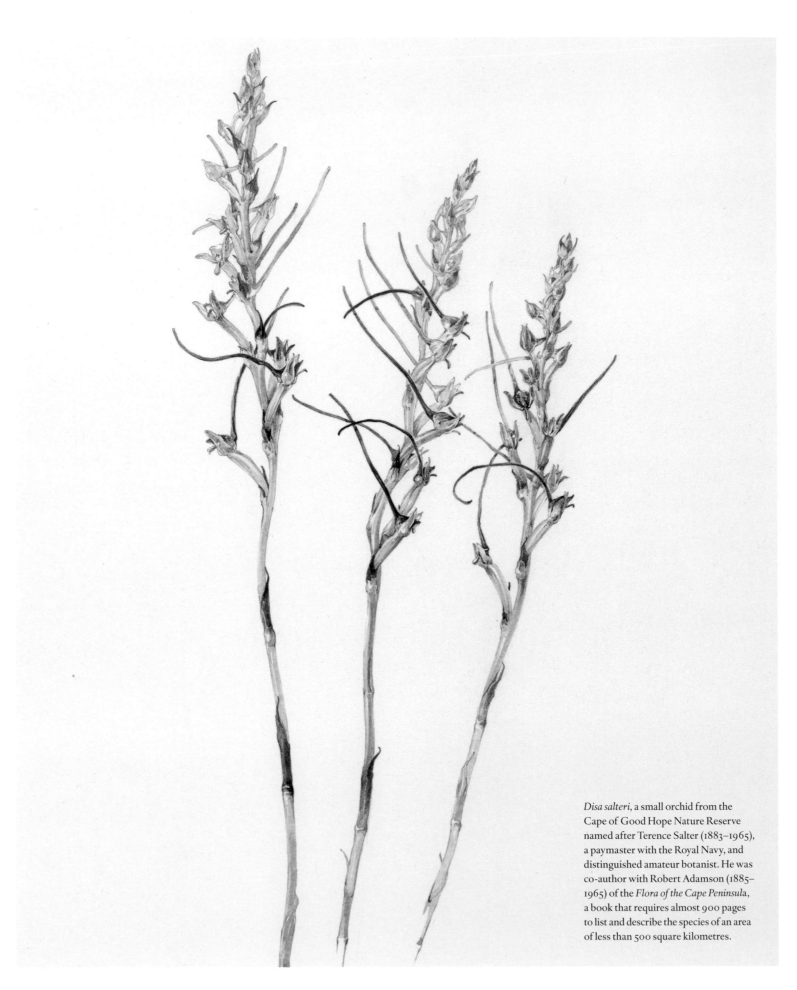

Disa salteri, a small orchid from the
Cape of Good Hope Nature Reserve
named after Terence Salter (1883–1965),
a paymaster with the Royal Navy, and
distinguished amateur botanist. He was
co-author with Robert Adamson (1885–
1965) of the *Flora of the Cape Peninsula*,
a book that requires almost 900 pages
to list and describe the species of an area
of less than 500 square kilometres.

References & Further Reading

The starting, and in many cases, finishing points for the background to many of the plant collectors are to be found in Mary Gunn and Leslie Codd's indispensable *Botanical Exploration of Southern Africa*. Conrad Lighton's *Cape Floral Kingdom* is also a very useful and readable source of information.

In researching the history, collection and cultivation of the Cape flora, I would not have got very far without the notable and prolific publications of Frank Bradlow, Neville Brown, Graham Duncan, Neil du Plessis, Vernon Forbes, Dick Geary-Cooke, Peter Goldblatt, Mia Karsten, John Manning, Charles Nelson, Ted Oliver, Rowland Raven-Hart, John Rourke, Peter Slingsby and Ernst van Jaarsveld.

In addition to journal and magazine articles too numerous to mention, the following is a selection of books that can be referred to for further details of the history of the Cape and its plants at home and abroad.

HISTORY AND BIOGRAPHY

AXELSON, E. (ed). (1988). *Dias and his Successors.* Saayman & Webb, Cape Town.

BEINART, W. (2008). *The Rise of Conservation in South Africa. Settlers, Livestock, and the Environment 1770-1950.* Oxford University Press, Oxford.

BOONZAIER, E., Malherbe, C., Berens, P. and Smith, A. (1996). *The Cape Herders. A History of the Khoikhoi of Southern Africa.* David Philip, Cape Town & Johannesburg.

BRADLOW, F. R. (1994). *Francis Masson's Account of Three Journeys at the Cape of Good Hope 1772–1775.* Tablecloth Press, Cape Town.

BURROWS, E. H. (1988). *Overberg Odyssey. A Chronicle of People and Places in the South Western Districts of the Cape.* Swellendam Trust, Swellendam.

BURROWS, E. H. (1994). *Overberg Odyssey. People, Roads and Early Days.* Swellendam Trust, Swellendam.

COOMBE, E. and Slingsby, P. (2000). *Place Names in the Cape.* Baarsdkeerder, Muizenberg.

COMPTON, R. H. (1965). *Kirstenbosch, Garden for a Nation.* Tafelberg, Cape Town.

CULLINAN, P. (1992). *Robert Jacob Gordon 1743–1795.* Struik Winchester, Cape Town.

DESMOND, R. (1977). *Dictionary of British and Irish Botanists and Horticulturists.* Taylor & Francis Ltd, London.

DESMOND, R. (2007). *The History of the Royal Botanic Gardens Kew.* Kew Publishing, Kew.

ELLIOTT, B. (2001). *Flora – An Illustrated History of the Garden Flower.* Scriptum Editions, London.

FORBES, V. (1965). *Pioneer Travellers in South Africa.* A. A. Balkema, Cape Town & Amsterdam.

FORBES, V. S. (ed). (1986). *Carl Peter Thunberg Travels at the Cape of Good Hope 1772–1775.* Van Riebeeck Society, Cape Town.

FORBES, V. S. and Rourke, J. (1980). *Paterson's Cape Travels 1777 to 1779.* Brenthurst Press, Johannesburg.

GREEN, L. G. (1951). *Tavern of the Seas.* Lawrence Timmins, Cape Town.

GRIBBIN, M. and Gribbin, J. (2008). *Flower Hunters.* Oxford University Press, Oxford.

GUNN, M. and Codd, L. E. (1981). *Botanical Exploration of Southern Africa.* A. A. Balkema, Cape Town.

GUNN, M. and du Plessis, E. (1978). *The Flora Capensis of Jacob and Johann Philipp Breyne.* Brenthurst Press, Johannesburg.

HORWOOD, C. (2007). *Potted History – The Story of Plants in the Home.* Frances Lincoln, London.

HUTCHINSON, J. (1946). *A Botanist in Southern Africa.* Gawthorn, London.

KARSTEN, M. (1951). *The Old Company's Garden and its Superintendents.* Maskew Miller, Cape Town.

KNAPP, S. (2001). *Potted Histories.* Scriptum Editions, London.

KUTTEL, M. (1954). *Quadrilles and Konfyt – The Life and Journal of Hildagonda Duckitt.* Maskew Miller, Cape Town.

LINDSAY, A. (2005). *Seeds of Blood and Beauty – Scottish Plant Explorers.* Birlinn, Edinburgh.

LIGHTON, C. (1960). *Cape Floral Kingdom.* Juta, Cape Town.

MCCRACKEN, D. P. and McCracken, E. M. (1988). *The Way to Kirstenbosch.* National Botanical Gardens, Cape Town.

NORTH, M. (1993). *A Vision of Eden.* The Royal Botanic Gardens, Kew.

QUINTON, J. C. and Lewin, A. M. (eds). (1973). *François Le Vaillant, Traveller in South Africa.* Library of Parliament, Cape Town.

RAVEN-HART, R. (1967). *Before Van Riebeeck.* Struik, Cape Town.

RAVEN-HART, R. (1971). *Cape Good Hope 1652–1702. The first 50 years of Dutch colonisation as seen by callers.* A. A. Balkema, Cape Town.

ROOKMAAKER, L. C. (1989). *The Zoological Exploration of Southern Africa 1650–1790.* A. A. Balkema, Rotterdam.

SLINGSBY, P. and Johns, A. (2009). *T. P. Stokoe, the man, the myths, the flowers.* Baardskeerder, Muizenberg.

THOM, J. (ed). (1952–58). *Journal of Jan van Riebeeck.* Van Riebeeck Society, Cape Town.

WARNER, B. and Rourke, J. (1996). *Flora Herscheliana.* Brenthurst Press, Johannesburg.

WILLSON, E. J. (1961). *James Lee and the Vineyard Nursery, Hammersmith.* Hammersmith Local History group, London.

YELD, J. (2004). *Mountains in the Sea – Table Mountain to Cape Point.* South African National Parks, Constantia.

BOTANY AND HORTICULTURE

BAKER, H. A. and Oliver, E. G. H. 1967. *Ericas in Southern Africa.* Purnell, Cape Town.

BROWN, N. and Duncan, G. 2006. *Grow Fynbos Plants.* South African National Biodiversity Institute, Cape Town.

BROWN, N. Jamieson, H. and Botha, P. (1998). *Grow Restios.* National Botanical Institute, Kirstenbosch.

BROWN, N. Kotze, D. and Botha, P. (1998). *Grow Proteas.* National Botanical Institute, Kirstenbosch.

BRUYNS, P. V. (2005). *Stapelias of Southern Africa and Madagascar*. Umdaus Press, Hatfield.

COOMBS, S. V. (1936). *South African Plants for American Gardens*. Frederick Stokes, New York.

CROUS, H. and Duncan, G. (2006). *Grow Disas*. South African National Biodiversity Institute, Cape Town.

DIBLEY, R. and Dibley, G. (1995). *Streptocarpus*. Cassell Educational Ltd for the Royal Horticultural Society, Wisley.

DUNCAN, G. (1998). *Grow Agapanthus*. National Botanical Institute, Kirstenbosch.

DUNCAN, G. (2000). *Grow Bulbs*. National Botanical Institute, Kirstenbosch.

DUNCAN, G. (2002). *Grow Nerines*. National Botanical Institute, Kirstenbosch.

DUNCAN, G. (1988). *The Lachenalia Handbook*. Kirstenbosch Botanic Gardens, Cape Town.

DU PLESSIS, N., and Duncan, G. (1989). *Bulbous Plants of Southern Africa*. Tafelberg, Cape Town.

ELIOVSON, S. (1956). *South African Wild Flowers for the Garden*. Howard Timmins, Cape Town.

FAIRBRIDGE, D. (1924). *Gardens of South Africa*. Maskew Miller, Cape Town.

GOLDBLATT, P. (1986). *The Moraeas of Southern Africa*. National Botanic Gardens, Cape Town.

GOLDBLATT, J. (1998). *The Genus Watsonia*. National Botanic Gardens, Cape Town.

GOLDBLATT, P. and Manning, J. (1998). *Gladiolus in Southern Africa*. Fernwood Press, Cape Town.

GOLDBLATT, P. and Manning. J. (2000). *Cape Plants. A Conspectus of the Cape Flora of South Africa*. National Botanical Institute of South Africa, Kirstenbosch.

GOLDBLATT, P. Manning, J. and Dunlop, G. (2004). *Crocosmia and Chasmanthe*. Timber Press, Portland.

HILLIARD, O. M., and Burtt, B. L. (1991). *Dierama – The Hairbells of Africa*. Acorn Books, Johannesburg.

KEY, H. (2003). *1001 Pelargoniums*. Batsford, London.

MANNING, J. (2007). *Field Guide to Fynbos*. Struik, Cape Town.

MANNING, J., Goldblatt, P. and Snijman, D. (2002). *The Color Encylopedia of Cape Bulbs*. Timber Press, Portland.

PATERSON-JONES, C. (2007). *Protea*. Struik, Cape Town.

REBELO, T. (1995). *Proteas. A Fieldguide to the Proteas of Southern Africa*. Fernwood Press, Cape Town.

ROURKE, J. P. (1982). *The Proteas of Southern Africa*. Centaur, Johannesburg.

SCHUMANN, D., Kirsten, G. and Oliver, E. G. H. (1992). *Ericas of South Africa*. Fernwood Press, Cape Town.

VAN DER WALT, Vorster, P. J. and Ward-Hillhorst, E. (1977–88). *Pelargoniums of Southern Africa* (3 vols). Juta, Cape Town/Purnell, Johannesburg/National Botanic Gardens, Kirstenbosch.

VAN JAARSVELD, E. (1994). *Gasterias of South Africa*. Fernwood Press, Cape Town.

VAN JAARSVELD, E. (2006). *The Southern African Plectranthus*. Fernwood Press, Cape Town.

VOGTS, M. and Paterson-Jones, C. (1982). *South Africa's Proteaceae, know them and grow them*. Struik, Cape Town.

WILKINSON, A. (2007). *The Passion for Pelargoniums. How they found their place in the garden*. Sutton Publishing, Stroud.

GENERAL NATURAL HISTORY

COWLING, R. and Pierce, S. (2009). *East of the Cape.* Fernwood Press, Cape Town.

COWLING, R. Richardson, D. and Paterson-Jones, C. (1995). *Fynbos, South Africa's Unique Floral Kingdom.* Fernwood Press, Cape Town.

FRASER, M. and McMahon, L. (1994). *Between Two Shores – Flora and Fauna of the Cape of Good Hope.* David Philip, Cape Town.

MANNING, J. and Paterson-Jones, C. (2007). *Field Guide to Fynbos.* Struik, Cape Town.

MCMAHON, L. and Fraser, M. (1988). *A Fynbos Year.* David Philip, Cape Town.

PARKER, R. and Lamba, B. (2009). *Renosterveld: A Wilderness Exposed.* LR Publishers, Cape Town.

PAUW, A. and Johnson, S. (1999). *Table Mountain, A Natural History.* Fernwood Press, Cape Town.

PATERSON-JONES, C. (1997). *The Cape Floral Kingdom.* Struik, Cape Town.

The Botanical Society of South Africa's excellent wildflower guides combine to cover the entire Cape Floral Kingdom, with individual volumes dedicated to popular and flower-rich areas such as Table Mountain and the Cape Peninsula, the West Coast, and Namaqualand. See:
www.botanicalsociety.org.za/publications/wildflowerguides.php

USEFUL WEBSITES

Botanical Society of South Africa *www.botanicalsociety.org.za*

Cape biodiversity hotspots *www.biodiversityhotspots.org/xp/hotspots/cape_floristic*

South African National Biodiversity Institute – includes Kirstenbosch National Botanical Gardens. *www.sanbi.org*

PlantzAfrica – information on indigenous plants. *www.plantzafrica.com*

The Biodiversity and Wine Initiative – promotion of a sustainable and environmentally-friendly wine industry at the Cape. *www.bwi.co.za*

Kogelberg Biosphere Reserve, a major conservation area near Cape Town. *www.kogelbergbiospherereserve.co.za*

South African National Parks *www.sanparks.org*

Cape orchids *www.capeorchids.co.za*

WWF South Africa *www.wwf.org.za*

Indigenous Bulb Association of South Africa *www.safricanbulbs.org.za*

Suppliers of Cape bulbs and seeds *www.africanbulbs.com, www.capeseedandbulb.com, www.silverhillseeds.co.za*

Index

Index of plants and animals, selected place names, and major characters
Text reference, *Illustration*, **Illustration by Liz Fraser**

First published in 2011 by
Royal Botanic Gardens, Kew
Richmond, Surrey, TW9 3AB, UK

www.kew.org

ISBN 978 1 84246 389 5

British Library Cataloguing in Publication Data
A catalogue record for this book is available from the British Library.

Front cover illustration: Liz Fraser
Authors' photograph: © Ken Paterson

For further information on the project visit www.thesmallestkingdom.co.uk

Edited by Alison Rix
Proofread by Michelle Payne
Design, typesetting and page layout by Lyn Davies Design

Printed and bound by Firmengruppe APPL, aprinta druck
Wemding, Germany

For information or to purchase all Kew titles please visit
www.kewbooks.com or email publishing@kew.org

Kew's mission is to inspire and deliver science-based plant conservation
worldwide, enhancing the quality of life.

Kew receives half of its running costs from Government through
the Department for Environment, Food and Rural Affairs (Defra).
All other funding needed to support Kew's vital work comes from
members, foundations, donors and commercial activities including
book sales.